PROTECTION FOR + FROM HUMANITY

This page intentionally left blank.

ISBN 978-0-578-62298-9

PROTECTION for + from HUMANITY

3nd edition

Library of Congress
101 Independence Ave SE
Washington DC, DC 20540

Published by

SINQUANON
Los Angeles, CA 90210

Oversight: Nick Fox
Author: Todd Fox
Editor: E. Robert
Cover Design: J. Leick
Consult: Zanino
Review: M. Jones
Base: Context ARTS

Printed in the **United States of America**.

This publication is available in e-format.

Additional publications by this author are available. Visit *TourProtection.com* for more information.

"Viam inveniam aut faciam."

-Hannibal

DEDICATION

This book is dedicated to my father, **Nick Fox**. He remains the most influential person in my life. A rare and good man, who dedicated himself to his family, business, charity, and the outdoors. He was a self-made man with limited formal education who would rise from poverty through perspiration and inspiration. He was, and is, a stellar example for all mankind. It would be an understatement of epic proportions to say that I was lucky to have him as my father. I will never be able to fill his shoes, although I will never stop trying to reach the bar he set for me. His unparalleled commitment to me was the greatest gift of my lifetime.

FOREWORD & ENDORSEMENTS

Todd Fox presents his third publication with *Protection for + from Humanity*, a comprehensive work for security practitioners, protection specialists, law enforcement officers and those that are transitioning into protective service operations. It further provides a great source of knowledge for private citizens looking for proven concepts and systems to protect their respective environments.

Protection for + from Humanity is written in Mr. Fox's inimitable style, blending formal training in the Marine Corps and law enforcement with practical application as a highly sought-after and successful protective operator. Todd's passion, commitment and sharing of knowledge in this industry is unmatched and is evident.

This volume incorporates both soft and hard skills as well as numerous principles and skill sets that security practitioners must understand and regularly practice in order to be successful. It is an essential collection of processes that are needed in an ever-changing security landscape.

As a retired Lieutenant Colonel and U.S. Army Ranger, I can attest to the author's detailed knowledge and training skills. I have successfully utilized numerous skills and principles covered in the book both domestically and internationally in Iraq over the course of the past five years. This book is an absolute must read that you will find yourself referencing frequently.

A. Ornoski
Lieutenant Colonel
U.S. Army Ranger (Ret.)
Vice President of Operations
OSAC Chairman, Baghdad, Iraq

"Mr. Fox has operated at the highest levels of international business. He has proven that his organization is capable of functioning freely while maintaining discretion in matters of great importance."

Supreme Court Justice
Identity Withheld

"Todd Fox is an exception. He can teach it and he can do it. I've seen him in action and I've seen him teaching. He carries a unique ability to have mastered both aspects. Read the book!"

Renzo Gracie
UFC Living Legend

"Mr. Fox is an extremely accomplished security Subject Matter Expert who has an acute insight as a consummate professional. His perception and vision of what true security is goes unmatched - anywhere - within the security industry. We support his tireless contributions to the industry."

R. Kaneiss
CEO, MAGNI Intl. Group
U.S. Navy SEAL

"The instruction Mr. Fox provided to our team was of great benefit. His knowledge of the subject matter, attention to detail, and skilled instruction are what make him an industry leader."

Special Agent B.W.
FBI, SWAT Team Leader

"Todd's knowledge and ability to modify techniques and strategies to fit the environment is impressive."

Capt. Kevin Davidson
Division Director, State Police ST-65

OBJECTIVE & DISCLAIMER

This book was created to introduce the reader to general concepts, systems, and tools used in the protection of clients, family, or self. It is intended to provide insight into the realm of the protective mind for the average, law abiding citizen. The content of this book will likely expose the reader to ideas and perspectives not commonly known outside the military and law enforcement. **It will not qualify an untrained person to perform or direct professional protective security operations. Further study, extensive protective training, and real-world hands-on experience are absolutely necessary and required prior to engaging in protective work.** This publication serves as a primer on the subject matter.

"No chess player has ever found, nor is any likely to find, a sure way of winning from the first move."

-David Galula

TABLE OF CONTENTS

Chapter **ONE**

INTRODUCTION

It has become increasingly popular in recent years to note that modern times are much safer than previous historical eras of extreme human violence. This may hold true when looking at raw data, pie charts, and specified graphs. However, none of that is relevant when you, a client or a loved one is involuntarily thrust into the chaos of a violent encounter. The only thing that matters is your will to survive and your ability to carry out the skills you have trained prior to the event. This book acts as a starting point in the protective process. It is an insurance policy of sorts. A policy we hope you never have to use. If you do, we hope the material in this book assists in keeping you and your loved ones safe.

Everyone needs protection. They always have. You need protection from the elements, protection from predators, and protection from diseases. You may need protection from the co-worker that's trying to get you fired. Many people even require protection from themselves! In the modern world, we don't have to worry about the saber-tooth tiger, we have to worry about a lot more. In the day and age of people walking around with ear-buds in and eyes fixed on their smart phone screen, we are almost completely detached and unaware of our surroundings.

It's hard to defeat a threat when you never saw it coming. Threats come in many forms and approach from many angles. Daily life can throw a barrage of dangers at you. Sometimes they are life-changing events and other times, by sheer luck, you manage to avoid a life-or-death incident. Only one promise can be made: *You too need protection*. However, just knowing you need protection doesn't cut it. You need to know how to put protective elements in place that will work for you. Fortunately, you're holding a book that can help. So, keep reading.

Don't be afraid. Be prepared.

The survival instinct is the most important mechanism genetically hard wired into the human brain. Without it, we would cease to exist as individuals and as a species. Most people understand this on a visceral level, yet few people take the time to understand what can be done to improve their chances of survival in situations that present danger. Some simple concepts and applied strategies known by a few select groups can drastically improve survival odds in less than ideal situations. These critical concepts and proven strategies are the same ones that work for protective specialists, tactical law enforcement teams, and military special operations groups. More importantly, they work for average people. Moms, dads, grandmas, grandpas and everyone else. This special knowledge can be applied in day to day life and in unique situations like an overseas business trip or family vacation. The information in this book can be used in the real world, right now. In fact, our team of professionals are using it across the globe as you read this. If you're a parent or a business owner, your choices can often determine the outcome of serious events. You do NOT need to overhaul your life to make it safer. Prepared is not paranoid, it's an intelligent choice.

Whether you're talking left or right politics, war or climate change, both the cause and solution are the same...Humanity. When talking about the protection of a human, it is essential to understand both the cause of danger and the solution specific to the danger presented. Understanding the how and why is required prior to the solution being implemented or attempted.

The intent of this book is to cover key elements of protection that are not covered in my previous publications. Here you will find the conceptual side of protection that leads to the technical implementation. Many of the subjects crossover each other. The concepts in this book are essential to creating a plan unique to your life and/or business. It will provide you the necessary tools to develop critical thinking as it pertains to human safety.

Protective ability is an acquired skill. It requires intense practice and training ad nauseam to be effective. It requires the ability to think on your feet. It requires that you make specific decisions that will determine the outcome of a given situation.

There is no one single way that works 100% of the time. Our method is A way, not THE way. Here you will find suggested methods that we have found to work in most cases, most of the time. There are a thousand ways to skin a cat, but all start with a sharp knife. Consider the book you are holding a sharpening stone.

Keep an open mind.

Crawl, Walk, Run.

"Nature has made up her mind that what cannot defend itself shall not be defended."

-Ralph Waldo Emerson

Chapter **TWO**

SECURITY vs PROTECTION

These words are often interchangeable for most people. Even among people working in the industry, they say they provide security services as often as they say they provide protective services. The title of this book uses the word ***"Protection"*** over ***"Security"*** intentionally. The word protection suits the intent and contextual elements associated with the tactics, techniques, and procedures addressed in this publication. When protection is discussed in future chapters, it is most often done in reference to human protection. In the majority of cases these systems can be applied to the protection of places and things as well.

Below are the actual definitions of the words. There are clear differences in tone. It may be nuanced from the outside, but it is exceptionally relevant in the realm of protection.

SECURITY

Se*cu*ri*ty
/si'kyŏoritē
noun

1. ***The state of being free from danger or threat.***
   (Does this even exist?)

2. *The safety of a state or organization against criminal activity such as terrorism, theft, or espionage.*

PROTECTION

Pro*tec*tion
/prə'tekSHən
noun

1. ***The action of protecting someone or something, or the state of being protected.***
   (This is the objective.)

2. *A person or thing that prevents someone or something from suffering harm or injury.*

"The will to survive is instinctual; the ability to survive is LEARNED."

-USMC

Chapter **THREE**

MILITARY/LE vs PROTECTION

Many people transitioning from the military or law enforcement move toward the protective field. There is a major distinction between having served in the military or having worked in law enforcement and providing protection. They are NOT synonymous unless you have received specialized training and worked in the field. Below are some examples that demonstrate the major differences between the military and law enforcement mindsets and that of a Protection Mindset.

MILITARY & LAW ENFORCEMENT	vs	PROTECTION
Possess Official Authority		Have NO Authority
Plan Attack/Offense		Plan Defense
Pursue Subjects		Are being Pursued
Move to the Fight		Evacuate from the Fight
Vast Tools & Resources		Limited Tools & Resources

The list above addresses just a few of the obvious differences in operations. It's easy to be lulled into feeling comfortable when you transition from Mil/LE to Protection. It is, however, essential to start with an open mind.

"What lies behind us and what lies before us are tiny matters compared to what ***LIES WITHIN US****."*

-Emerson

Chapter **FOUR**

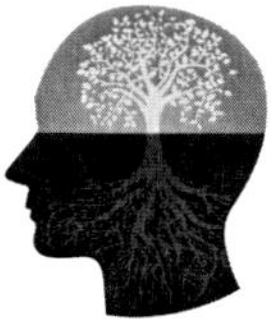

MINDSET

Mindset is critical in all high-stakes events. It is oftentimes the single most important factor in determining the outcome of a given event.

"If you think you can, or you think you can't...you're right."

This quote summarizes the power of belief and confidence, also known as your **mindset**.

Most people are capable of maintaining a positive mindset when things are going well within their controlled environment. This requires no special capabilities. When we refer to mindset, we are looking at the mindset required to survive in less than ideal, austere conditions. Some mental components are wired into a person's DNA. Others are developed at a young age due to personal circumstances. Many require specialized training to learn how to deal with, operate in and around, or survive/thrive in chaos or unstructured, non-permissive environments.

The only person who can decide *what* and *how* you think is **YOU**.

The number one weapon any person can access in a dangerous situation is their **BRAIN**. Even armed attackers need to rely heavily on their brain first before being able to effectively utilize their tool(s) or machine(s) as a weapon.

Lt.Col. Cooper once said that "*Owning a handgun doesn't make you armed any more than owning a guitar makes you a musician.*"

Your primary weapon, the human brain, is capable of detecting and processing subtle non-verbal cues that are advance indicators of dangerous situations before they actually unfold. This capability affords time *(and potentially space)*, which generally allow for better survival, or response options. At a minimum it offers an advantage over the alternative. A single second is invaluable in a lethal confrontation. In short, your **BRAIN** and associated **MINDSET,** give you the edge, or tactical advantage.

One mental lesson that has repeatedly presented itself across generations in the military, law enforcement, and the martial arts is this:

EVERY TIME YOU QUIT, IT BECOMES EASIER TO QUIT THE NEXT TIME. EVERY TIME YOU ENDURE (RESIST/FIGHT), IT BECOMES EASIER TO ENDURE THE NEXT TIME. = MINDSET

Recent scientific studies have shown that mentally quitting or enduring often translates into different realms and areas of life. There are dedicated fields in psychology that focus on this specific area. Modern day professional athletes often use performance psychologists to develop a better, more competitive edge. These specialized psychologists have the responsibility of helping individuals or groups of people identify the **positive mindset** for developing, enhancing, and maintaining optimal human performance in a variety of applications.

Some key areas to consider for individual mindset development are Visualization, Objective/Mission Focus, Endurance, Restraint, and Relaxation Under Pressure.

Remember that things are not what they are. They are what we think they are. Psychological value is the best kind.

The American Psychological Association (APA) offers a number of publications on the subject.

You can do a lot to test your mental limits while simultaneously building mental strength. Some easy areas to start include long fasts, cold showers, sleep deprivation, and extreme isolation.

The material provided in this book is designed to provide mental resources *(tools)* that will assist the reader in withstanding conflict, or adversity.

COOPER'S COLOR CODES

Lt.Col. Jeff Cooper developed a system to identify different states of mind. He did this in the form of a color code. This system was used to identify states of mental readiness. The system pre-dated current governmental color codes and shares no relational value.

WHITE – ***Unaware & Unprepared***
Condition White is the weakest state of mind. Survival odds are very low. It is equivalent to being in a day dream. The only chance of survival is at the hands of a clumsy, unskilled adversary. *Heart rate 60-80 BPM.*

YELLOW – ***Relaxed Alert***
Condition Yellow is no specific threat, but a preparedness of mind that acknowledges a chance an attack could occur. Saying to yourself *"I may have to do something to defend myself today."* Most people can operate in this condition with no added stress in their day to day lives. *Heart rate 60-80 BPM.*

ORANGE – ***Specific Alert***

Condition Orange is a specific alert. Something is on your radar. Your focus is on that potential threat. Saying to yourself, *"If "X" happens, I will do "Y" in response."* This condition can be maintained for substantial periods of time. *Heart rate 60/80-120 BPM.*

RED – ***Fight Mode***

Condition Red occurs when the aforementioned "X" happens and you are in the fight. *Heart rate 120/150-180 BPM.*

*The **U.S. Marine Corps** uses one additional color in their system that Cooper did not have.*

BLACK - ***Panic Mode***

Condition Black is a reference to being completely overwhelmed. This state would refer to the Freeze in Fight, Flight, or Freeze mode. This is a useless, undesirable mental mode in any combat engagement. *HR over 180 BPM.*

In short, Lt.Col. Cooper's color code system considers levels of mental preparedness which are designed to help people think before and during attacks.

"All men dream, but not equally. Those who dream by night in the dusty recesses of their minds, wake in the day to find that it was vanity: but the dreamers of the day are dangerous men, for they may act on their dreams with open eyes, to make them possible."

-T.E. Lawrence

Chapter **FIVE**

PROBLEM, ANALYSIS, SOLUTION

Protection can often be reduced to awareness and problem solving. Problems require solutions. Solutions generally require analysis. This process is very similar to the scientific method.

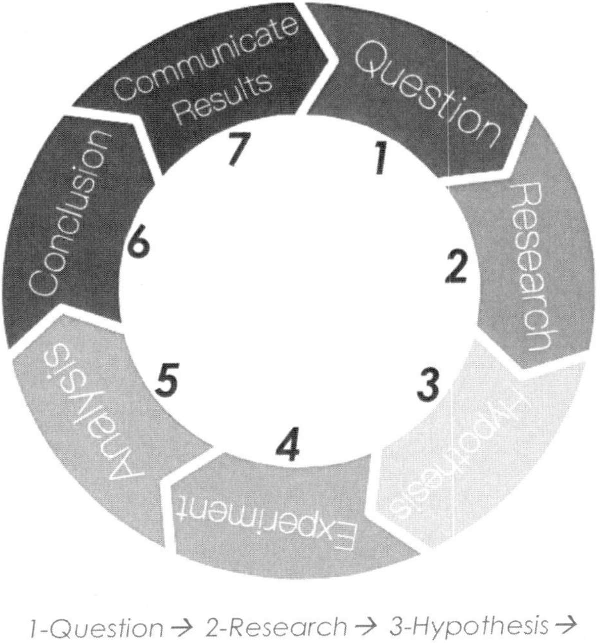

1-Question → 2-Research → 3-Hypothesis → 4-Experiment → 5-Analysis → 6-Conclusion → 7-Communicate Results to Protection Team

"If I had an hour to solve a problem, I would spend 55 minutes defining the problem and five minutes thinking about solutions."

-Albert Einstein

Einstein offers a good example of the importance of understanding the core of a problem. However, in protective realms, you will rarely be afforded the time to contemplate the problem in depth while it is unfolding. The thought will need to have occurred beforehand in training, though life experiences, or through contextual understanding. Reading this book would fall into Einstein's *"55 minutes"* category.

If you don't have time for the scientific method, reduce it down to ***ACT****.*

ASSESS THE SITUATION 

CREATE A PLAN 

TAKE ACTION 

A - *Take in all the information available.*
C - *Create a plan based on your abilities.*
T - *Implement the plan. Update as needed.*

A key component in a **realistic** plan is context.

CONTEXT

One of the most underestimated and crucial elements of protection is **<u>context</u>**. Context is often the key to true, deep understanding of a person or situation.

Context can be many things.

-Circumstances
-Surroundings
-Environments
-Backgrounds
-Scenes
-Atmospheres
-Factors
-Conditions
-Situations
-Settings
-Backdrops
-Climates
-Ambience
-Feelings

Understanding associated context allows you to fully process the totality of the circumstances. It provides you with better information for decision making. Utilizing context ultimately results in a higher probability of making the right, or most appropriate decision. In the world of protection, context can allow you to understand who the client actually trusts, an environmental influencer's point of view, what medical issues a co-worker has, or the feelings of your boss' child on a subject that may be essential to your success. Context is a decoder of sorts. It should be sought out in all situations that involve humans and safety. **Context provides insight and assists in filling in informational gaps.**

Robert Sapolsky, author of *BEHAVE: The Biology of Humans at Our Best and Worst,* suggests that it's helpful to understand human behavior as emanating from a series of concentric circles that provide multiple layers of CONTEXT. That's an important statement that should be considered throughout this book.

When discussing a problem with a foreign diplomat overseas, a single sentence put an hour-long conversation into context. It allowed me to properly digest what was available in the moment. It was a very simple statement. ***"Truth is not a matter of majority."*** *(Ministro Alberto P.D.)*

"Every morning in Africa, a gazelle wakes up. It knows it must run faster than the fastest lion or it will be killed. Every morning in Africa, a lion wakes up. It knows it must outrun the slowest gazelle or it will starve to death. It doesn't matter whether you are a lion or a gazelle. When the sun comes up, you'd better be running."

-Christopher McDougall

Chapter **SIX**

SITUATIONAL AWARENESS

It is impossible to protect anything or anyone without having an awareness of the situation that surrounds you. We will address some of the key areas of situational awareness individually in later chapters. While not listed as such, baseline/ anomaly identification, normalcy bias, contextual development, criminal minds, common dangers, required knowledge and response options are all part of situational awareness.

It has become exceptionally challenging for the modern citizen to maintain situational awareness. This is primarily due to smart phones, social media and shortened attention spans. Phones and social media could reasonably be considered *"attention suckers."* If you are fixated on your phone, you are missing what is happening around you. This usually occurs without a cost. In the wrong situation, this lack of awareness can be very costly.

Position yourself in a way that allows for both observation and defense or departure. This is not a decision based on paranoia. It is simply a low-energy method of preparation. Simple decisions save lives. This applies to many environments for many people. It may be "Don't cross the street without looking both ways" or "Don't put your back to the door." These are different techniques, but employed for the same purpose.

Time is another factor of awareness. You should endeavor to develop the habit of regularly looking at your watch when departing and arriving. It's also good to mentally note the time when significant events begin and end.

Intuition is also an area worthy of consideration. Your eyes and ears are fixed immediately outside your brain for a reason. This proximity shortens the transmission time in sensing elements subconsciously that are critical survival sensations. Don't ignore these information pulses. **TRUST YOUR GUT.** There are countless stories of this single awareness factor saving human lives.

"Complacency kills" is a statement made daily in jobs fields that deal with shooting, fighting, and driving at high rates of speed. It's a true statement that jibes with *"familiarity breeds contempt."* This is where the concept of normalcy bias seeps into our daily lives. We don't focus attention on things that we do often. This is particularly dangerous in high-stakes situations when we are mentally or physically fatigued. We expect things to fall into place as they usually do. **It is a bad day when high stakes/risk events, fatigue, and complacency come together.** Don't let it happen. Actively work against it in all realms.

IF/THEN THINKING

If/Then thinking is exactly what it sounds like. It's a method of keeping yourself in a relaxed state of awareness which allows you to objectively look at your environment and make decisions in advance. It can be done in an almost game-like manner.

What could happen to me today in this environment? If that happened, how could I solve the problem? Where could I go? What could I do?

In equation format it would look like

IF "**X**" HAPPENS TODAY…I WILL DO "**Y**" IN RESPONSE.

*"When I'm walking in the parking lot tonight, I will be consciously aware of my environment. If someone jumps out and attacks me before I get to my car (the "**X**") then I will hit them with the bag in my left hand and tase them with the taser in my right hand (the "**Y**")"*

One critically important thing to understand is that the decision must be made prior to any incident. This shortens the process by at least one human mental response time.

COOPER'S PRINCIPLES

Lt. Colonel Jeff Cooper, a former US Marine, was a leader in the field of self-protection and firearms. He was an innovative thinker who was able to transfer knowledge like no other person of his era. He is best known as the founder of the first American gunfighting school, *Gunsite*, almost fifty years ago. One of his publications, ***Principles of Personal Defense***, is timeless. It should be read by anyone who cares about their personal safety or that of their client or family. Lt. Col. Cooper presents seven core principals to be learned and applied in the defense of one's self or others.

PRINCIPLE ONE: **Alertness**

It's easy to lose awareness in the day and age of smart phones, tablets, and other mobile devices. This puts most of us at an immediate disadvantage. These attention sucking devices force us to lose our alertness. The exchange is at the cost of our personal safety in many cases, as we do not see things until it is too late to prevent them.

PRINCIPLE TWO: **Decisiveness**

A common saying in the military and protective realms is *"Hesitation kills!"* This has repeatedly been proven in combat and by science. Using the Recognition Primed Decision Making and If/Then thinking concepts will allow for more decisive actions in critical moments.

PRINCIPLE THREE: **Aggressiveness**

The sad truth is that modern times have brainwashed people into believing aggressiveness is a bad trait. When applied in the appropriate context or environment, it may be the only solution that allows for survival. This is forgotten in times of prosperity and wealth, but will always be understood in extreme environments. Survival can be extreme.

PRINCIPLE FOUR: **Speed**

Moving and thinking fast will often solve problems. If you can beat your adversary through the thinking cycle, you will likely beat him in the physical realm. In a basic sense, you can imagine two combatants fighting in close range. One has his weapon out, one holstered. Assuming all else is equal, the first one with sights aligned properly, and a smooth trigger pull, will likely win the fight. John Boyd's OODA Loop system, in the next chapter, will address this process further.

PRINCIPLE FIVE: **Coolness**
In this setting coolness could translate to calmness or smoothness. This feeling or vibe will translate to efficiency of motion/action. The smoother the motion is, the higher the probability the action will be effective. A common military saying is *"smooth is fast"*. It could easily be said that cool is fast.

PRINCIPLE SIX: **Ruthlessness**
This is another principle which is not understood during prosperous times. Lt. Col. Cooper makes a very potent statement at the beginning of this chapter. He says, *"Anyone who willfully and maliciously attacks another without sufficient cause deserves no consideration.*" The focus here is on the value of human life…yours, your family's, and/or your client's. The concept is that staying alive will often require actions that are ruthless to fully and completely stop an adversary. This requires you to be tough, even if it seems harsh from the outside or in hindsight.

PRINCIPLE SEVEN: **Surprise**
In any realm of strategy, business or combat, *surprise* is a key principle that provides an overwhelming advantage to the individual or group deploying it. This element is coveted by both criminals and law-abiding citizens. In fact, many strategies are predicated on this element alone. Surprise puts the other person/group on their heels and creates a slower, reactionary response, when deployed properly.

NOTE: The brief summary above does not do justice to Lt. Col. Cooper's teachings found in the actual publication. It is strongly recommended that you read *Principles of Personal Defense*.

720 CONCEPT

It's difficult to wrap your mind around the areas (*space*) that you are responsible for when protecting someone. Some people consider a 180-degree plane. Others says 360-degrees around the protectee. We typically speak in terms of 720-degrees. While the number is irrelevant, the intention is to make protectors think about the circle around them (360) and the circle above and below them (360). Things fall from buildings. Pot holes exist on streets. Cars don't always stop at red lights. All of those can injure or kill your client, family, or friend. Awareness should not have a special limit. In protection we frequently speak of having your ***"head on a swivel."*** This is a core concept that you should adopt in your daily life.

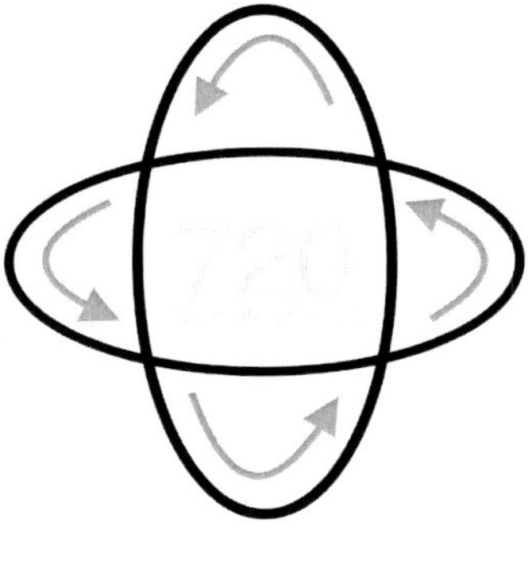

720 PROTECTION
CONCEPT
SNIPER IN
ELEVATED POSITION
HOLE IN
GROUND
FALLING DEBRIS
BAISC 360
PROTECTION
POISONOUS GAS
RADIATING FROM
BELOW GROUND

"It is not the strongest of the species that survives, nor the most intelligent that survives. It is the one that is the most adaptable to change."

-Charles Darwin

Chapter SEVEN

ADAPTABILITY & FORESIGHT

The ability to adapt to the situation, environment, or culture is essential to success in protection. In extreme cases, it can make the difference between living and dying. You will be required to perform certain tasks that you have never performed before. In most cases, the problems you trained for will present themselves in different ways. This can make the task considerably more challenging. The trick is to understand that plans will change and things will go wrong. Expect for things to veer off course and be ready to adapt. Try to stay flexible and relaxed when possible.

Always create options and continue to look for alternative methods. The Marine Corps nailed this concept when they developed the mantra of ***Improvise, Adapt, and Overcome***. You must wrap your head around this before you step foot out the door. This doesn't mean that because things are likely to change, you don't need a plan. It means you need multiple plans and the ability to think on your feet and create a new plan as the situation unfolds. **The situation dictates the tactics**. It is essential to understand that adapt or die is a reality in the world of protection.

As they say, change brings opportunity. It also brings danger. Be ready to adapt to changing environments.

CONTINGENCY PLANNING

Most people understand that predicting something before it happens is an uncommon and invaluable skill. This skill is part and parcel to the protection of a person, place, or thing. It is most akin to professional prudence. These types of skills are commonly taught to military and law enforcement personnel and frequently used to prevent disasters.

A number of time-tested, globally proven systems exist for this particular area.

Conceptually, a layered approach is always best in protective operations. A good starting point for this is the planning acronym of PACE.

P - Primary Plan
A - Alternate Plan
C - Contingency Plan
E - Emergency Plan

This is as simple as it seems. PACE essentially translates to having three back up plans. While this is not common in day-to-day life, and may even be considered excessive by some, it's an invaluable mind tool for volatile circumstances or environments. If nothing else, it is a reminder of options that require forethought.

EXAMPLE:

WEAPONS	COMMUNICATIONS
P - Pistol	**P** - Issued Radio
A - Rifle	**A** - Issued Cell Phone
C - Knife	**C** - Satellite Phone
E - Hands	**E** - Smoke Signals

DECISION MAKING

Learning expedient systems for decision making also falls within the foresight category. The acronym OODA appropriately covers this in the protective field.

In the 1960's Col. John Boyd, a U.S. Air Force fighter pilot, developed the OODA loop cycle. It was a simple way to understand the decision-making process/cycle in aerial combat. Typically, the fighter pilot who mentally cycled through the OODA loop the fastest won the fight. Today the system is used in many sectors.

OODA is an acronym that stands for *Observe, Orient, Decide, Act*.

Observe the situation.
Orient yourself to it.
Decide your best course of action.
Act on your plan.

Each time a variable in the fight equation changes, the OODA Loop restarts. This continues until the fight is over.

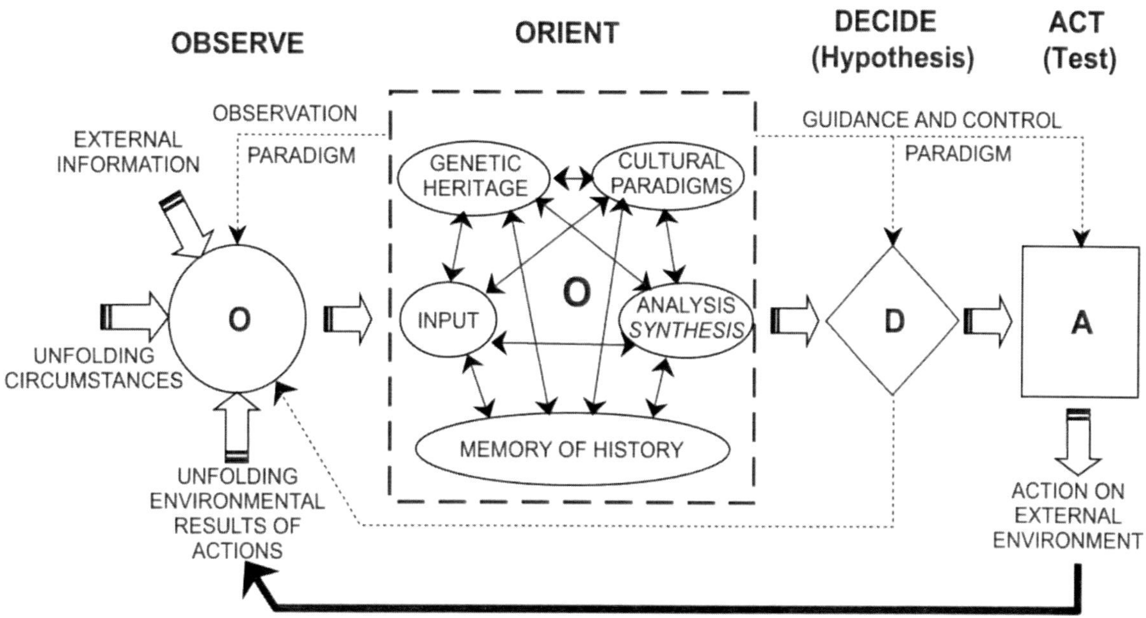

Source: Based on Boyd (1996)

"What is predictable is preventable."

-Gordon Graham

Chapter **EIGHT**

Environmental Analysis

It is important, in any environment, to understand what is normal activity, behavior, etc. In protection this is referred to as a BASELINE. The primary reason to identify a baseline is to be able to know when something outside the norm is occurring. This is referred to as an ANOMALY. An anomaly can be above or below the baseline. It can be benign or critical. The standard rule of thumb is that if you have an anomaly from the baseline, a decision is required. This often equates to a change in behavior, action, or movement.

If you take daily walks down your neighborhood street at the same time, you likely know what the baseline is for that area. If you see an 80-year-old man sitting alone on his porch every day for months, this would become part of your baseline. If you walked by and saw five young men with him, that would be an anomaly above the baseline, as there has been something added that was not normally there. Additionally, you could have an anomaly below the baseline where the man is not on his porch. Both should produce the same question. Why has this changed? How could this affect me? Should I change behavior?

EQUATION:

Baseline + Anomaly = DECISION

In simple terms, the **decision** forced by the anomaly commonly plays out in one of three ways.

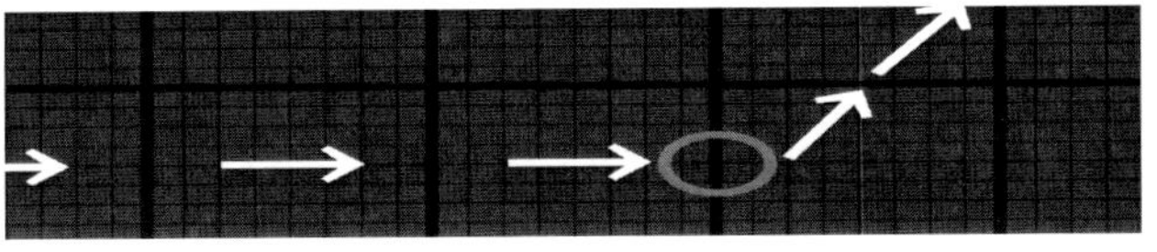

1. ***CONTINUE*** *as planned*
2. ***CHANGE*** *plan, route, activity*
3. ***CANCEL*** *operation, job, movement*

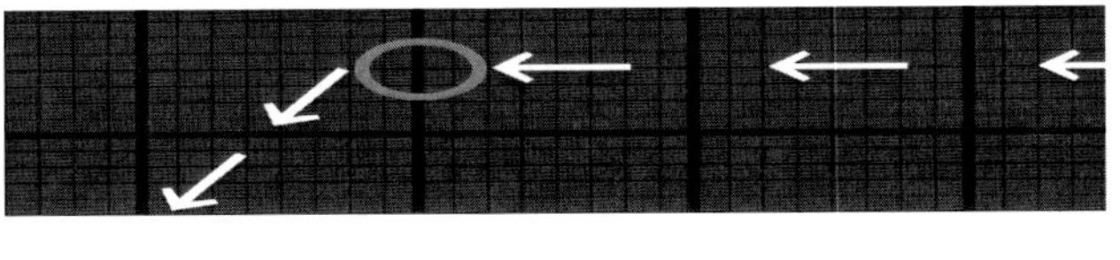

NORMALCY BIAS

Protective operations require the understanding of normalcy bias to prevent falling victim to it. According to psychologists, normalcy bias (cognitive bias) is *"a bias to believe that things will always function the way things normally function"*. In other words, the natural human tendency to try to fit anomalies into something understandable and routine. This often leads people to disbelieve or minimize threats, or warnings. This is a dangerous possibility for people tasked with protection.

RECOGNITION PRIMED DECISION MAKING

One way to improve your reaction is by planning/training for it. In the RPDM Model (Klein), the decision maker will pick up cues and indicators that lead them to recognize patterns. Based on these patterns and the decisions they have to make; the person chooses a single course of action they believe will assist in achieving the desired outcome. In RPDM, the decisionmaker comes up with an "action script" for the most common problems or likely scenarios. This allows them to quickly execute the decision when time is essential. As protection professionals develop their field craft, their ability to implement RPDM successfully improves as they are better able to correctly recognize the key features of a problem **and see/select appropriate solutions.**

There are an endless number of possibilities when it comes to survival scenarios. Having already looked at **RPDM** and **IF/THEN** thinking, the next step is to look at which types of events you should be most concerned with as a team, company, or individual. What areas are most important for time use, energy expenditure, and training? How can policies and procedures (TTPs) be implemented to support this area?

Gordon Graham is the co-founder of Lexipol and a retired LEO who is best known as a risk management expert. Mr. Graham is one of the few humans who maintains the special ability to extrapolate small environmental details and translate them into exceptionally funny and relevant stories that drive the point home.

While he maintains a wealth of knowledge, his most famous model is that of the Risk vs Frequency quadrant. In this process Mr. Graham looks at four specific areas and how they impact decision and severity of outcome *(consequence)*.

According to Mr. Graham, his risk/frequency analysis system is used to *"identify the specific tasks that have the highest probability of ending up in a tragedy."* He then advocates that teams/agencies create and implement control measures in places to prevent these tragedies from occurring. This should be done in policies, procedures, tactics, techniques.

Specialized training should FOCUS on the **High Risk/Low Frequency** quadrant, as this is where most events fall apart and have the greatest consequences. In particular, those events with **non-discretionary time**, or NDT.

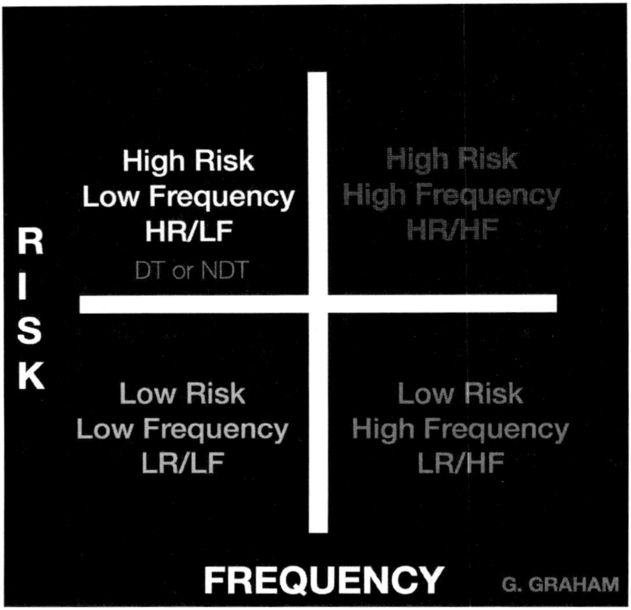

High Risk/Low Frequency Events have serious consequences, but happen so infrequently that most people/agencies do not have the skills to deal with the issue. A HR/LF event that has "DT" or *Discretionary Time* that affords the ability to think through the problem offers a much higher probability of success or resolution. Conversely, HR/LF events with "NDT" or *Non-Discretionary Time* are the most likely of all types of events to go wrong and have horrible consequences. Think of NDT as a physical, mental, or emotional AMBUSH of sorts. It occurs so quickly that you have seconds to decide your response and implement it appropriately. Conversely, DT is something that could potentially be solved in hours or days.

HR/LF: *Gunshot Wound, Stabbing*

High Risk/High Frequency Events have serious consequences, but happen so frequently that most people/agencies do have the necessary skills or training to deal with the issue.

HR/HF: *Heart Attack, Asthma Attack*

Low Risk/Low Frequency Events have no significant consequences and occur infrequently, so time, energy, and training are not often focused on these events.

LR/LF: *Broken Finger/Toe, Chicken Pox*

Low Risk/High Frequency Events have no significant consequences and occur frequently, so most are aware of how to solve them. As they produce little to no consequence, most have little time or concern for these areas.

LR/HF: *Trip, Fall, Cut, Bruise*

"When faced with just one opponent and you oppose yourself...you're already outnumbered."

-Dan Millman

Chapter **NINE**

CRIMINAL MINDS

Most people have been a victim of crime at some point in their lives. Understanding how criminals think and operate is fundamental to protection. Of particular value is how criminals choose their targets. Most criminals follow certain patterns of thought and behavior. The first thing they need is something of value. This can be as simple as a watch, wallet, car, or any material possession believed to be of value that is owned by the target. In the case of kidnapping, the individual themselves hold value, provided someone or some group is willing to pay for their release. Criminals come in many forms. Fortunately, most of them seek out the same elements in a target. They prefer a weak target (victim) who is in possession of a valuable asset(s), and is not likely to fight back. They'd also like it to be as fast as possible with as few people as possible witnessing the crime. This affords the criminal the greatest chance of success while limiting their chances of being caught by police or adversarial observers.

In the most basic sense, the best option is to make yourself invisible or useless to the criminal. In protection, we call this **target hardening**.

1) **DON'T STAND OUT**
   Blend into the crowd. Don't wear highly visible/valuable clothing, jewelry, bags, or other items that draw attention within that realm. Follow the local or regional customs and attempt to be "***GREY***".

2) **KNOW WHAT THE PROBLEMS CAN BE AND WHERE THEY CAN COME FROM**
   It is much easier to defeat an adversary if you know where they can come from and how they may attack you. This requires forethought and constantly changes when moving through different environments.

3) **HAVE A WAY OUT**

Know where you can go to from any location or position. Routes *(foot and vehicle)*, safe havens, directional awareness, and maps should be part of your core emergency plan. This plan too will change when passing through different environments.

4) **LOOK LIKE YOU WON'T GIVE UP, AND DON'T GIVE UP IF YOU ARE ATTACKED**

If it is suspected that a criminal has selected you and is about to make his move, make sure he feels like you're going to fight back. More importantly, have the ability to fight back. If you can't escape and are forced to fight, give it everything. It may be the fight of your life. Literally.

DON'T FEED CRIMINALS

Don't knowingly provide criminals with the information needed to do a target assessment on you or your family! *Don't understand?* The next page will offer context and drive the point home.

WHAT AM I TELLING THEM?

1 - *"Baby on Board"* tells criminals that you will have your hands full and a limited ability to deal with an attacker.

2 - *"Honor Student @ Sea Cliff Elementary"* tells criminals where your most valuable possession can be found M-F, 8:00am-3:00pm.

3 – *"I Love My Poodle"* tells criminals that you have a small, non-threatening, non-guard dog at home.

4 – *"Edward Jones Employee Parking Pass"* tells criminals where you work.

5 – *"Proud Army Wife"* tells criminals that your husband spends long period of time away from the house.

6 – *"Hunter"* tells criminals that you likely have a nice gun collection at home.

TERRORIST OPERATIONS

One very special type of criminal is the terrorist. Terrorist activities are generally ideologically based and target innocent civilians. The objective is to instill fear into the population. Terrorist groups range from untrained children to complex networks of well-trained, well-funded threat actors. The best way to deal with these groups is through early detection or identification. While this can occur at any point prior to the attack, it is most likely to happen during pre-attack surveillance inside the PLANNING phase of the attack cycle. In the realm of terrorism, the cycle is a loop. It generally starts with the selection of a target, then planning *(which includes surveillance and weapons acquisition),* then deployment of tactical operators, then the attack. In terrorism, the attack is then followed with an exploitation phase to ensure maximization of the damage created.

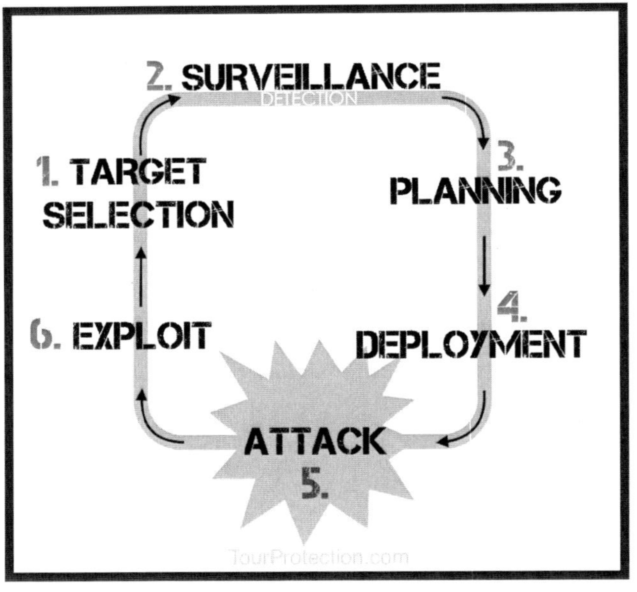
2. SURVEILLANCE
DETECTION
1. TARGET
SELECTION
3.
PLANNING
4.
DEPLOYMENT
6. EXPLOIT
ATTACK
5.
TourProtection.com

COMMON PRE-ATTACK INDICATORS

In almost all cases, indicators are present prior to attacks. The Baseline Anomaly process is one way to identify what does not fit into the environment. Here are some other common indicators.

- *Statements made in person or via other platforms about the intent and/or methods of future attacks.*
- *Individuals who are visibly uncomfortable in a low/no stress environment.*
- *Individuals asking questions that don't fit the environment or scenario.*
- *Individuals evading direct questions regarding a simple subject.*
- *Individuals wearing inappropriate clothing for environmental conditions.*
- *Theft of uniforms, badges, official vehicles, access/gate openers.*
- *Testing/probing of security structure.*
- *Small groups of individuals communicating in unusual ways.*
- *Weapons concealed or visibly presented in a common interaction.*

*"One size does **NOT** fit all."*

-Experience

Chapter **TEN**

COMMON DANGERS

Danger drives decisions. The first question to ask is ***what am I protecting my client, family member, or friend from?*** This is the foundation of every other element in this equation. This will drive your decisions and actions. If done properly, and quickly, you will likely be successful. If you misidentify this portion, you will likely fail.

What is the protector's job? The simple answer is to keep the person you're protecting safe at all times. Now, the more important question is *"safe from what?"* This is where most people start going in the wrong direction. Protectors often place most of their attention in the wrong area. Misplaced or wasted attention can lead to a significantly lower probability of prevention, detection, and successful response. There are five general areas you need to protect against.

1) PHYSICAL VIOLENCE

This is the most obvious area, which includes the standard attacks, such as being punched, kicked, choked, thrown, bludgeoned, slashed, stabbed, shot, etc. This is often a point of fixation for protectors, but actually occurs the least. If you fixate on this, you will likely miss one of the other four areas.

2) ENVIRONMENTAL/UNINTENDED HAZARDS

Most people have had that unfortunate experience of falling off of something, running into something, or having something fall on top of them. These are the environmental or unintentional hazards of daily life. Luckily, in most cases this doesn't end in death. These hazards are usually preventable through situational awareness alone.

3) REPUTATIONAL DAMAGE

This type of risk is normally informational in nature. It frequently stems from a specific act or action, which can be public or private, real or perceived. It may present serious financial and legal ramifications. In some cases, it may even lead to loss of revenue, career, bankruptcy, and/or community standing. Many of these cases are preventable through specific, intentional measures put into place to reduce a given exposure in advance. A good example of this for parents is cyber bullying via social media, which has resulted in a significant number of teen suicides.

4) LOSS OF PERSONAL INFORMATION

Like reputational damage, this exposes the victim to a number of potential attacks that may result in long-term financial damage. This too is preventable through verbal discretion and physical protective measures. This can be as simple as safeguarding personal credit cards with a foil type barrier that prevents walk-by criminals from extracting information electronically. Protection can also be in the form of an insurance policy that covers this type of attack. Specific countermeasures can be in put place to combat the effects of almost any unauthorized informational breach.

5) ANYTHING *(That could negatively impact you, your client, a family member, or a friend.)*

This catch-all is never ending. It means that the job of protecting a person, place, or thing is never really done. It's ongoing and dynamic, and that means it requires a lot of attention. This is where it becomes evident that protective skills are actually needed as much by parents as they are for a person working in the protection business.

You must know *what the problems can be* and *where they can come from* in order to protect any person, place, or thing. This is possibly the most important element of the protection equation. If you lack situational awareness, you are not likely to be successful in protecting yourself, or someone else.

"Protective Services are an absolute ZERO FAIL mission. Protection professionals don't get a pause, reset, or do-over button."

G. Breed III
Sr. Instructor
USG PSO

Chapter ELEVEN

REQUIRED KNOWLEDGE & ABILITIES

Developing the **protective mind** is really about understanding the responsibilities of the protector. RESPONSIBILITY is the *ability to respond*. As the protector, you are tasked with solving problems. These problems often happen without notice and when least convenient or completely unexpected.

To properly respond to the problem, you should:

1) Know what POTENTIAL PROBLEMS could occur and where they could come from.

In order to prevent something from occurring, you must know (*be aware*) that it can occur in the first place. For example, a child may decide to touch a hot stove because they are curious and have no idea what the consequences will be. You must seek to understand what problems may be presented in a given environment before you get there. This requires constant assessment of the environment. We will discuss specific methods for this later in the book.

2) See the POSSIBLE SOLUTIONS to the problems.

All problems have solutions. Knowing the problem doesn't help unless you can see potential solutions to that specific problem. Having related previous experience is the best tool. The next best is creativity.

3) Be able to implement the CORRECT RESPONSE, under stress.

Some solutions fit well, others create more problems, and yet others are simply inappropriate. You need to select the correct solution/response to/for your problem based on the totality of the circumstances. Once you have done that, you must be able to implement that in the real world under significant physical, mental, and emotional stress. In lethal situations, you only have to be right 100% of the time. If you fail, you will never have to worry about solving that problem, or any other, again. It's important to understand that responding requires advance thought and continuous mental and physical preparations.

These steps must occur in sequential order, often in seconds. Missing any of the above areas will result in failure. A common theme here is that training in advance will help to solve the problem quicker and with better results.

"Simplicity is the ultimate sophistication."

-Da Vinci

Chapter **TWELVE**

RESPONSE OPTIONS

It's always a good idea to have or create options. Sometimes doing nothing is the best option. Sometimes it's not. When evaluating exposure to risks or threats, consider the three primary response options listed below.

1. BEST OPTION

Don't be where danger commonly occurs!

If you know that an area consistently presents danger, stay out of that area when at all possible. This is the definition of being proactive in the security realm. Through directly limiting your exposure, you are strategically implementing risk reduction. If you know there will be a protest outside an embassy down the street at 12:00 pm today, avoid any travel or movement in that direction. If you know a particular bar has a history of fights, don't go to that bar if you don't want to potentially be targeted.

2. GOOD OPTION

Leave at the first sign of a problem.

For protectors, this area is not always intuitive. It's important to understand that staying and fighting is the last resort. If you find yourself in trouble, leave as soon as you can. Getting yourself, your family, or clients out as soon as possible can save your life and theirs. This is commonly referred to in the military as "Getting off the X." As soon as you see or feel something is not right, move. Even if you are not the direct target of the attack, you don't want to become a target of opportunity.

3. LAST/BAD OPTION

Stay and fight. (counterassault)

Unfortunately, this is a reactive option. The adversary has chosen the time, place, and method of attack. The only element you control here is your *response* to the attack. It is important to understand that staying and fighting means every other option has failed. This is commonly misunderstood in the security field. It needs to be well understood in protection. Fighting allows you the potential to stop or eliminate the threat. From a tactical perspective it also affords you an opportunity to create distance between you and the threat. This ultimately gives you the best chance of survival. In this particular area it's important to select who will fight and who will withdraw well in advance. This is true in both protective environments and with families. Like all areas of protection, this must be drilled repeatedly for it to translate under stress.

In the event that multiple protectors are covering a person or place, the general protective theory on use of assets is this;

MINIMUM to the THREAT,
***MAXIMUM** to the **PRINCIPAL**.*

That means that while multiple people are covering down on the principal, or acting as human meat shields, only one may be directly engaging the threat. As with all tactics, this decision will be driven by the format of the attack, available response options, and overall context of the situation.

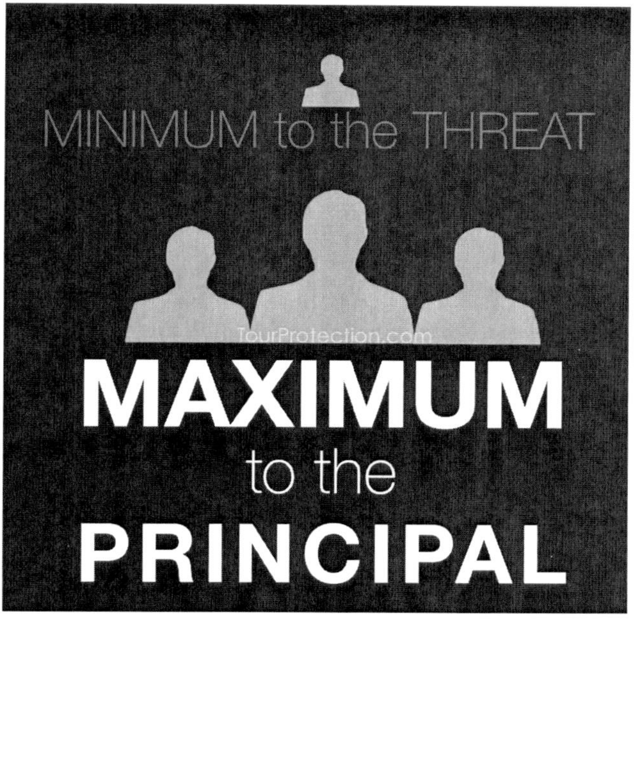
MINIMUM to the THREAT
TourProtection.com
MAXIMUM
to the
PRINCIPAL

"In the realm of ideas everything depends on enthusiasm...in the real world all rests on perseverance."

-Goethe

Chapter **THIRTEEN**

ATTACK & COUNTER EQUATIONS

Once the criminal mind and modus operandi are understood, elements of the attack can then be analyzed. There are four primary areas to focus on when breaking down attacks. It's critical to understand which elements the attacker selects/controls and which elements the target selects/controls.

The ATTACKER selects

1) **Location** of Attack
2) **Time** of Attack
3) **Method** of Attack

The TARGET selects

4) **Response** to Attack

Note that the attacker controls 75% of the variables in this equation. The response is a **REACTION**. Action almost always beats reaction. The exception is when responses have been planned and properly trained in advance of the attack. Train the way you want to fight.

ATTACKER SELECTS

1. Time
2. Location
3. Method

TARGET SELECTS

4. Response

COUNTERASSAULT

The intent of fighting back, or countering the attack, is to **preserve life**. This can playout through direct threat elimination or through the suppression of the attack. The downside of suppressing an attack is that it is a temporary solution to the problem. This response affords time to the individual or group being attacked. The time can be used to disengage from the attack and create space/distance between the attacker(s) and the target(s). Unfortunately, the target rarely gets to select if elimination or suppression occurs. This is usually dictated by the spatial environment, skills of those involved, relevant tool/assets available, and number of attackers and targets involved.

COUNTERASSAULT EQUATION:

The closer the attacker, the less skill is required to do harm to the target.

CLOSER = **More Dangerous**

COUNTERASSAULT = **Time** + **Distance**

In general, attackers close or take space and defenders create space or distance. The greater the spatial distance between the attacker and target, the higher the PROBABILITY of **SURVIVAL** for the target.

Proxemics, or the study of spatial distances between individuals, will be addressed later in the book.

*"Someone told me that for each equation I included in the book, it would **halve** the total sales."*

-Stephen Hawking

Chapter **FOURTEEN**

EQUATION MANIPULATION *(DISRUPTION)*

In the practice of protection, it is always better to prevent a negative action in advance of the attempt. However, that can be difficult to achieve if you are operating alone or have limited access to intelligence or assets. In this case attack **disruption** is more probable than prevention.

Math is not something that is synonymous with protective operations. In fact, most working in the field probably have a distain for it when presented in a classic format. Manipulating an equation *or situation* means that you rearrange/restructure/re-sort the equation *or situation*. This is, actually, a type of math. There are two general approaches that can be used in equation manipulation (EQM). Both are used to change or **disrupt** the adversary's intended objective *(outcome)* in a given situation or attack. In a broad and general sense, this can be viewed as protective math.

One of the processes requires focusing on a specific action that you do **NOT** want to happen in a given time and place. This is often referred to as the "X". The X is in reference to the time and place of a successful attack. As in, X marks the spot (*of the attack*). A common statement in the protection world is to ***"Get off the X."*** *Meaning get away from the place the attack is likely to happen.* ***The "X" represents the negative outcome of the situation (equation).***

If A+B+C=X, you would add or remove variables from the left side of the equation so that your outcome is never **X**. In other words, anything but X is an acceptable outcome. X**-1** is ok. X**+1** is ok.

EXAMPLE BELOW

John likes to party, but he can't afford to get into any more trouble. He works long hours Monday through Friday, so he can only go out on the weekend.

Friday night he takes his kids for ice cream. He runs errands and spends time with the family during the day on Saturday and Sunday. His favorite local bar is Bubba's Tavern. On Friday nights, he puts the kids to bed at 9:00pm. He likes to meet his friends Bob and Joe for drinks around 10:00pm at Bubba's Tavern. John recently had a legal issue and can't afford to have any more, as he is now on probation. Bubba's is known for its bar fights, especially during late night hours on weekends.

Here is what the protective math would look like to keep John out of trouble.

Any outcome but X is acceptable.

EQUATION A+B+C=X
A=Friday
B=Time after 11:00pm
C=Bubba's Tavern
X=Bar Fight *(or high probability of attack)*

MANIPULATION OPTIONS
Removal of **A**,
by going on a Monday = Outcome ***W***
Removal of **B**,
by going at 7:00pm = Outcome ***Y***

Removal of **C**,
by skipping the tavern = Outcome **Z**
Addition of Police to the tavern = Outcome **V**

Any of these manipulations would change the outcome from X to a non-X, acceptable outcome.

The second process is used to manipulate the variables so that the outcome of the equation results in **"Y"** ONLY. This means any other outcome other than "Y" equates to adversarial success, or your failure.

If A+B+C=X, you would add or remove variables in the equation until you achieved the outcome of Y.

It's 2014 and Gunny Killian is a Marine who is deployed to Northern Iraq to fight ISIS. His platoon is directed to do a patrol in the area of the Gogjali Market in Mosul later in the afternoon. An advance team is sent out to perform reconnaissance on the area. Upon arrival, they immediately notice that the market was operating at only 10% of the normal level. Most doors and windows were closed in the area surrounding the market. Pedestrian foot traffic was almost non-existent.

EXAMPLE BELOW

The only acceptable outcome is Y,
successful attack not possible.

EQUATION A+B+C=X
A= Mosul, Iraq 2014
B= Gogjali Market
C= U.S. Marine in Uniform
X= Successful Attack
(or high probability of successful, lethal attack)

MANIPULATION OPTION
Delaying the patrol and remaining inside the compound in Erbil = Outcome **Y**

These examples are a bit silly, but they should drive home the point of *equation manipulation* or *disruption* as applied to the protection of humans. Math is not heavily used in protection but, contextually, this single area should be known in theory and practiced in the field.

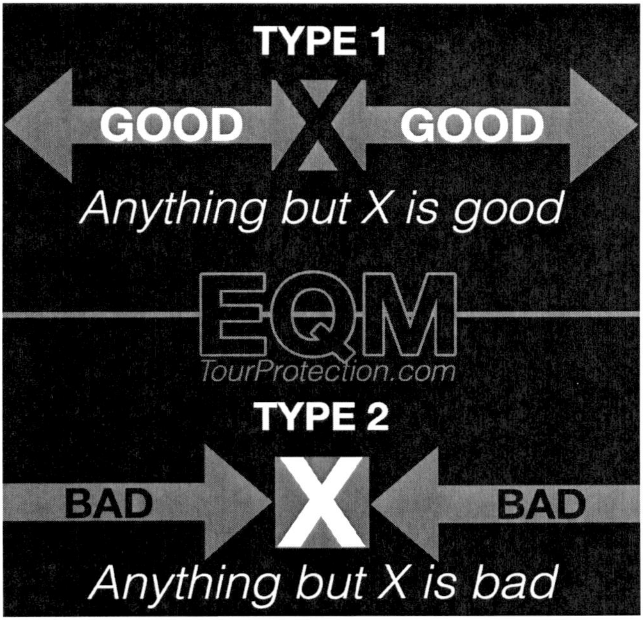
TYPE 1
GOOD
X
GOOD
Anything but X is good
EQM
TourProtection.com
TYPE 2
BAD
X
BAD
Anything but X is bad

*"Focus on the outcome,
not the obstacle."*

-Unknown

Chapter FIFTEEN

STRESS & INOCULATORS

Stress affects everyone on earth. Yet most don't understand it or manage it well. What does the human body do in a high-stress, life or death situations? This is commonly referred to as **FIGHT**, **FLIGHT**, or **FREEZE** mode. This is not a daily occurrence for most people. However, it should be understood by anyone working to protect someone or something of value. If you understand what the body's response may be or what it could feel like, you will likely manage it better.

1. **AUDITORY EXCLUSION**
   Auditory Exclusion is the loss of some or all hearing in a high stress situation. It is typically triggered by a hormonal heart rate spike (180+BPM). This is thought to be wired into the human body to remove the distraction of sound in a potentially lethal encounter.

Something such as a baby crying, or loved one screaming could become a deadly distraction.

2. **TUNNEL VISION**
Tunnel Vision is the loss of peripheral vision resulting in a tunnel like view. This physiological response to stress is also experienced following a spike in hormonal heart rate (180+BPM). This response is also thought to be wired into the human body to remove the distraction of sights in a potentially lethal encounter. Seeing a loved one hurt may distract a person from the immediate threat at hand, ultimately leading to their demise.

3. **POOLING OF BLOOD AWAY FROM ARMS/LEGS**
In extreme stress the blood may leave both arms and legs and pool toward the center of the chest. This is thought to be a hard-wired measure of protection to retain the blood should you lose an arm or leg in a confrontation. This results in less

immediate blood loss and increases the chance of survival.

4. **LOSS OF FINE MOTOR SKILLS**
   In extreme stress, the loss of fine motor skills is common. One reason is the aforementioned pooling of blood away from extremities. This leaves little blood in the fingers and toes. It is believed that this also assists or forces the individual to make large, gross motor skill movements which are more likely to defeat a lethal threat.

5. **HORMONAL HEART RATE**
   An immediate spike in heart rate based on fear is known as hormonal heart rate. It happens under extreme stress and is commonly referred to as an "*adrenaline dump.*" It is not the same as standard sports induced heart rate increases. The physiological change is instant and often takes a person from a resting heart rate of 60 BPM up to 180+BPM without physical activity involved.

Common high stress situations/causes include:

- *Attacks on Principals*
- *Attacks on Security Personnel*
- *Shootings*
- *Violent Physical Attacks*
- *Kidnappings*
- *Hostage Scenarios*
- *Vehicle Accidents*
- *Natural Disasters*

The best way to prepare for any type of stress is to expose yourself to a realistic but controlled, less dangerous, version of the incident(s) you are likely to face. Repeated training and exposures will generally reduce the body's physiological response to real incidents. Breathing is another critical element to reduce the negative effects of stress. It is the only element of an adrenaline dump that you can actively control. Slow, steady breathing reduces the heart rate, allows for better blood flood *(which allows the brain to have rational thoughts)* and assists in re-establishing semi-normal visual and auditory senses. A commonly taught practice in high-threat operations is called *"combat breathing.*" It is a basic and proven method of slowing human respiration. Systems vary, but the standard is as follows:

1) *Deep inhale over a four (4) second period/count.*
2) *Hold that breath, lungs full, for four (4) seconds.*
3) *Exhale for a four (4) second period/count.*
4) *Hold your breath, lungs empty, for a four (4) second period/count.*
5) *Repeat cycle.*

Results can be felt in as little as 60 seconds or approximately four iterations of this cycle.

The best approaches to address physiological stress include:

- *Creating realistic training plans*
- *Implementing slow and deep breathing training and awareness (combat breathing)*
- *Thinking rationally for options*
- *Maintaining positive thoughts*
- *Focusing on reasons to endure and overcome specific scenarios*

"Complexity is your enemy. Any fool can make something complicated. It is hard to make something simple."

-Sir Richard Branson

Chapter **SIXTEEN**

SOFT SKILLS vs HARD SKILLS

In the military, law enforcement and security contracting realms, skills are often broken down into two categories. Most are trained in both hard and soft skills with a core competency in only one. Hard skills generally refer to shooting, driving, fighting (DA-direct action) or defending (QRF-quick reaction force) and communicating. Soft skills include areas of human intelligence, including collection, analysis, facilitation, and communications. In the realm of protection both hard and soft skills are required to function smoothly and produce consistent results. Both skill sets rely on the other for support. Both skill sets have numerous cross-over points. Both skill sets are needed for operational harmony and mission accomplishment. Both should be studied and trained.

"Human behavior flows from three main sources: desire, emotion, and knowledge."

-Plato

Chapter SEVENTEEN

SOFT SKILLS

HUMAN BEHAVIOR

The most constant and accessible source of information in protection comes directly from humans in the operational environment. In the military this is known as human intelligence or HUMINT. One component of HUMINT is the behavior of a subject, or multiple subjects. Keep in mind that no two people are identical in behavior. Not all behavioral indicators are confirmation of intentional deception. Understanding the aforementioned caveats, most people's behavior can offer a relatively high degree of accuracy when taken in context with a baseline profile. The legal phrase for this is "*based on the totality of the circumstances.*" Timing and frequency of indicators are critical factors. The gathering of information/knowledge is known as the collection phase. This collected material allows the information assessment phase to begin. The cycle of collection and assessment are never ending and generally move at a rapid pace in modern times.

The next page lists several physical indicators that may be useful in assessing human behavior. The key areas of body posture, eye contact, and gestures are reviewed. These are only a few of the many physical indicators that exist. These are, however, very strong and consistent indicators.

BODY POSTURE

TRUTHFUL	UNTRUTHFUL
Open	*Closed Off*
Upright	*Slouches*
Frontally Aligned	*Angles Away*
Forward Leaning	*Pulls Back*
Normal Posture Changes	*No Changes or Frequent*

EYE CONTACT

TRUTHFUL	UNTRUTHFUL
Direct Eye Contact with Natural Breaks	*Avoids Eye Contact or Locks Eyes*

GESTURES

TRUTHFUL	UNTRUTHFUL
Limited Grooming Gestures and Limited Body Shifting	*Visible Grooming Gestures to Release Tension*

"Behavior is belief driven."

-Unknown

SPACIAL PERCEPTION & SIGNIFICANCE

We walk, stand, and talk next to other people throughout the day. It is something most people do without thinking too much about. But what happens when a person gets too close to you in line? What is your visceral reaction?

Proxemics is defined as the study of the cultural, behavioral, and sociological aspects of spatial distances between individuals. It also considers the effects of the physical distances on their interactions. It can be a component of nonverbal communications known as haptics, kinesics, vocalics, and chronemics. Spatial arrangements are often symbolic and unconscious. This means that proxemics can offer significant indicators, or clues, the observer(s). It may be used to better understand a subject's cultural origins, alignments or possible physical intentions.

It is important, for these purposes, to assess proxemics in a protective capacity. The four primary distances we will consider will be intimate, personal, social, and public. Each range presents information that can be utilized to better understand the situation or relationship between subjects or groups.

INTIMATE DISTANCE

Intimate distance is for embracing, touching, or whispering. This is typically reserved for the closest of friends or family and is within 6-18 inches.

PERSONAL DISTANCE

Personal distance for good friends is typically 2-4 feet.

SOCIAL DISTANCE

The social distance kept between most people interacting with acquaintances is typically 5-10 feet.

PUBLIC DISTANCE

The distance kept for public speaking is approximately 10-20 feet.

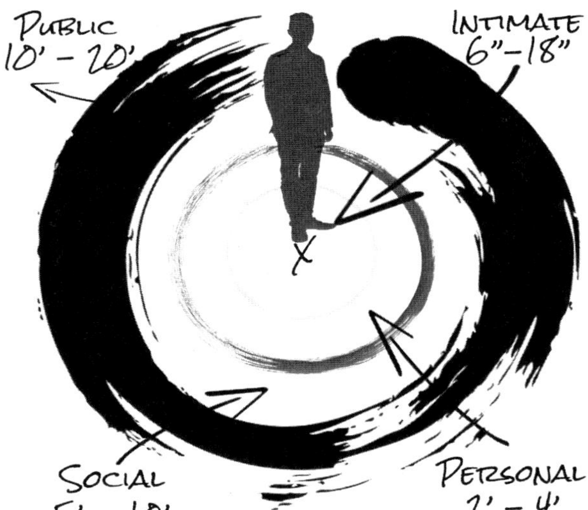

OBSERVATION CONTINUATION

Protection requires constant and continuous observation. ***"Head on a swivel"*** is a full-time function. It is also the technique that is most often responsible for environmental data collection. Roughly 90% of decisions made are done with the visual collection of data.

Human behavior is a central point of observation when attempting to gain perspectives. Watching for behaviors, patterns, and trends requires training coupled with intuition. When assessing the possibility of threat, the USMC Combat Hunter Program uses the acronym ALERT.

Alone & Nervous
Loose or Bulky Clothing
Excessive Fidgeting
Rigid Midsection
Tightened Hands

While ALERT is not all encompassing, these behavioral indicators are reliable when multiple elements are present together. *These behaviors require decisions.*

A Alone and Nervous.
Unresponsive, distant, and/or inattentive.
Appearing to be in disguise. Appearing in a trance.

L Loose or Bulky Clothing.
Inappropriate for season.
Exposed wires.

E Excessive Fidgeting.
Clock watching, scanning area, checking their "6".

R Rigid Midsection.
Rigid posture with minimal body movement, arms close to sides.

T Tightened Hands.
Repeatedly patting upper body.

A number of factors draw the attention of the human eye. These areas can be used to identify a subject or object to determine its relevancy or function in the environment. According to the U.S. Marine Corps' Combat Hunter Program, key areas include, but are not limited to the following areas:

Line, Edge, Outline

A boundary, border, separation or distinction. Usually against a contrasting background. The human eye picks these up easily.

Shape

A visual characteristic surface configuration of a thing. When an outline is perceived, it becomes a shape.

Contrast

A difference or dissimilarity between compared objects. Often seen as dark/light.

Color

Aspect of things based on light reflection.

Texture

Distinctive physical composition of structure of something. Respective to size, shape, arrangements of parts, and surface feel.

Rhythm & Flow

A pattern created by lines, forms and colors.

Movement

Greatest disturbance to environmental baseline. Easily detected with the human eye.

Light

Light impacts colors. Certain surfaces reflect light, while others absorb it. Light also casts shadows.

Passive/Negative Space

Positive Space takes up mass, solid objects. Generally, it's not possible to see through positive space. Negative Space is the space between two positive spaces.

APPEARANCE OF DISTANT OBJECTS

Gauging distances is a learned skill that requires practice. It can change in rural, suburban, and urban areas. It can also change between elevations such as flatlands and mountains.

Below is a general guideline for gauging distances when looking at the human body between 100 and 700 meters.

100m - *Human eye can distinguish one inch of detail. Can see facial features.*

200m - *Facial features can't be seen any longer. Skin color and equipment visible. Hairline usually visible.*

300m – *Clear body outline and skin tone visible. Hairline not visible.*

400m – *Body outline visible. Skin tone is no longer visible.*

500m – *Body tapers. No head and shoulder outline, blurs into blob. Light is still visible between legs.*

600m – *Body appears wedge shaped. No apparent head.*

700m – *Body outline now a solid wedge shape.*

Other areas of distance assessment are important as well. They can be developed in specific areas of operation.

Examples:

*-Average US car license plate is **1 foot** long*
*-Average US residential door is **3 feet** wide*
*-Average US car is **15 feet** long*
*-Average US school bus is **40 feet** long*
*-Average US commercial jet is **210 feet** long*
*-Average US football field is **330 feet** long*

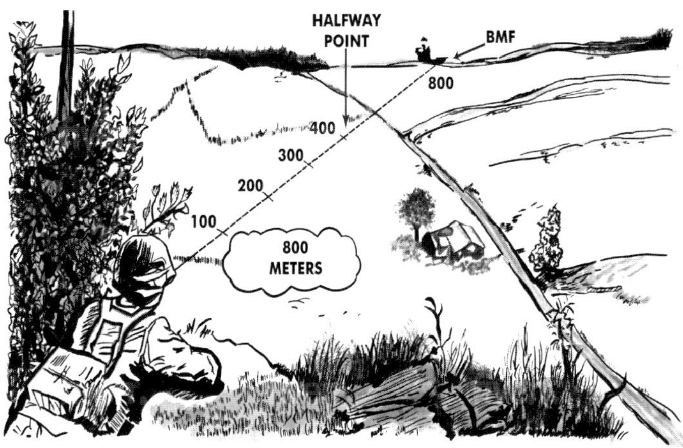

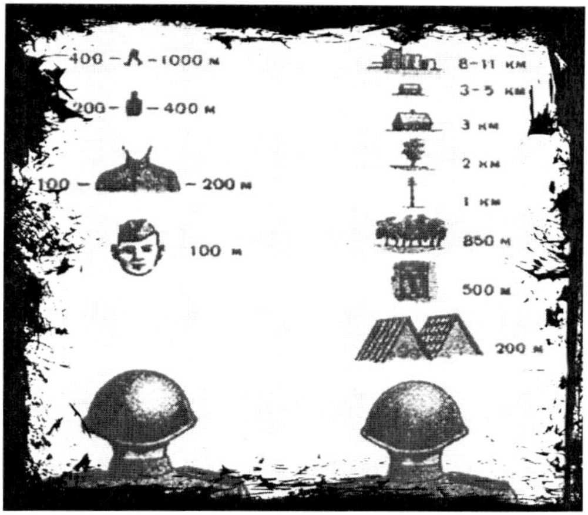

When traveling to new environments and time permits, work on your sketching skills, or tactical diagraming. Attempt to draw landmarks to scale. Note approximate distances based on a common or visible reference. When possible, check your work by walking off the actual distances, and measure items such as cars and houses.

1. Orientation (North Arrow)
2. Major Terrain Features
3. Sketch Building Outline
4. Sketch Windows & Doors (Size, Direction Opens)
5. Sketch Load-Bearing Wall Lines
6. Sketch Suspected Interior Walls
7. Sketch Trees, Shrubs, Plants
8. Sketch Fences & Vehicles (Type/Details)
9. Fine-tune Sketch
10. Add Team Positions & Approaches

"There are two ways to be fooled. One is to believe what isn't true. The other is to refuse to believe what is true."

-Unknown

Data, Info, Intel

Data, information, and intelligence are all a part of the same continuous sequence. They are, however, different in both definition and function. They are equally important for precise decision-making. The end objective is to secure intelligence that can be acted upon. This affords sound decision-making capabilities for the holder, which typically result in beneficial outcomes.

Data in its rawest form is a recorded certainty from a point in time. It's a snapshot of an event/action. That event/action might be a conversation, a transaction, or an interaction with something or someone.

Information is the bringing together of multiple data points. Information is created when you take data from an event/action and put it into a narrative format. It provides context out of the data.

Intelligence is the next stop once the information is vetted. Intelligence is the use of processed information to drive decisions. Information tells a story and intelligence paints a picture.

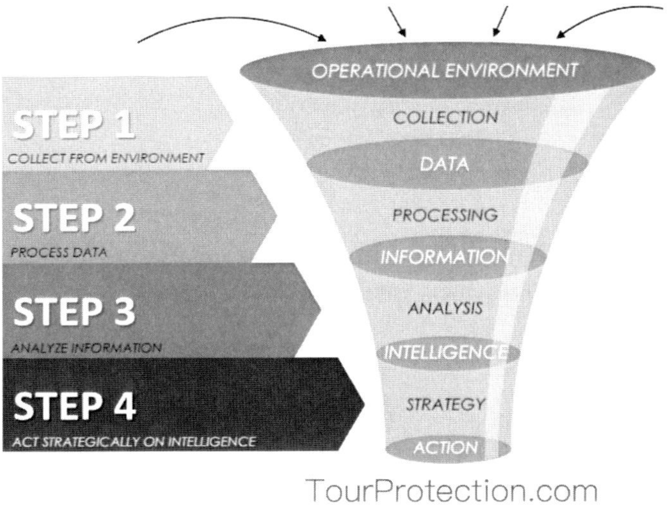

Source Reliability & Information Validity

Information is a critical component of the success of any person, organization, or operation. Getting information is only the starting point. Raw information alone is not enough. That information must be processed to ensure it is valid. Once this is confirmed, it is considered to then be "intelligence", which may become actionable. Two default areas that are considered in information assessment are the source's reliability and the validity of the information provided.

Both military and law enforcement use similar systems to assess information and sources. They are typically done in a grid format and assigned a value.

In the military a six (6) point scale is most often used to rate the source reliability and an eight (8) point scale is used for information validity. It is not uncommon to see systems that use three to five (3-5) point scales. Below is an open source example of the military format for source reliability and information validity.

SOURCE RELIABLITY *(A-F mil format)*

A-Reliable

No doubt of authenticity, trustworthiness or competency. Informant has a history of complete reliability.

B-Usually Reliable

Minor doubt about authenticity, trustworthiness, or competency. Has a history of valid information most of the time.

C-Fairly Reliable

Doubt of authenticity, trustworthiness or competency. Has provided valid information in the past.

D-Not Usually Reliable

Significant doubt of authenticity, trustworthiness, or competency. Has provided valid information in the past.

E-Unreliable

Lacking authenticity, trustworthiness, and competency. History of invalid information.

F-Cannot be judged

No basis exists for evaluating the validity of information.

INFORMATION VALIDITY *(1-8 mil format)*

1-Confirmed

Confirmed by other independent sources. Logical in itself. Consistent with other information on the subject.

2-Probably true

Not confirmed. Logical in itself. Consistent with other information on the subject.

3-Possibly true

Not confirmed. Reasonably logical in itself. Agrees with some other information on the subject.

4-Doubtfully true

Not confirmed. Possible but not logical. No other information on the subject.

5-Improbable

Not confirmed. Not logical in itself. Contradicted by other information on the subject.

6-Misinformation

Unintentionally false. Not logical in itself. Contradicted by other information on the subject. Confirmed by other independent sources.

7-Deception

Deliberately false. Contradicted by other information on the subject. Confirmed by other independent sources.

8-Cannot be judged

No basis for the evaluation of the validity of the information.

Below in grid format:

SOURCE RELIABILITY	
A	Reliable
B	Usually Reliable
C	Fairly Reliable
D	Not Usually Reliable
E	Unrealistic
F	Cannot be Judged

INFORMATION VALIDITY	
1	Confirmed
2	Probably True
3	Possibly True
4	Doubtfully True
5	Improbable
6	Misinformation
7	Deception
8	Cannot be Judged

HUMAN MOTIVES

It's important to understand what motivates people to act. This is true in both personal and professional capacities. In order to operate in a protective capacity, you have to know where threats can come from. This includes internal and external threats. Motives for human compromise should be a key point of awareness. This generally falls into one of four categories, which form the acronym MICE.

Money
Ideology
Coercion
Ego

These categories generally transcend nationality, culture, ethnicity, race, and gender. Specific individuals or groups will weigh more heavily certain categories. Identifying which matters most is critical to any action plan. These factors need to be considered for internal and external threats, risks, and vulnerabilities.

Money *is relatively obvious to most people. Greed is not an uncommon trait. One area that frequently drives this variable is personal debt. Conversely, some cultures do not value money in the same way the modern world does. Additionally, some people have more than enough money and the value of more is limited.*

Ideology *is a driving and consistent factor everywhere. This typically breaks down into two primary areas; religion and politics. Ideology is most often used in manipulating the subject. It is a common tool in false flag operations.*

Coercion *is the act of persuasion through force or threats. If you don't do "X", then we (threat actors) will do "Y".* Blackmail also falls into this category.

Ego *deals with a person's sense of self-importance. In this context, a manipulation or negative action occurs through the subject's ego being stroked or taken advantage of.*

"If I wanted to kill myself, I would climb up to your EGO level and jump to your IQ level."

-Napoleon

PRINCIPLES OF PERSUASION

Another, more modern, method to consider is that of Dr. Robert Cialdini's six principles of persuasion. While they do not replace MICE, these principles are becoming more commonly used in all business realms, including protection.

The six principles are:

Reciprocity
Authority
Scarcity
Consistency
Liking
Consensus

Reciprocity is the obligation of people to give back to others what they first have received from them.

Authority is the concept that people will follow the lead of people with knowledge and credibility.

Scarcity is of value because people strongly desire what is not easy to have or access.

Consistency is the conformity in the application of something. It is often associated with logic, accuracy, and fairness.

Liking is the feeling of fondness or special regard for someone.

Consensus is a general agreement among a group of people.

These principles are frequently used by online and social media platforms. They are specifically written into their coding with the intent to solicit a desired human response from the "*user*" or target. These principles are equally valuable when applied for the purpose of protective HUMINT. They can be applied individually, but are best used collectively, in concert. While these principles can be deployed at any time, they are particularly effective when used to build a network of human assets. These principles do not have borders and transcend all cultures. You should consider this system/concept a useful and proven social tool for protective operations both domestically and internationally.

"Doing more of the same and expecting different results is not a particularly effective strategy."

-Admiral Rogers

ASSESSMENTS & ANALYSIS

An assessment is a process used to identify potential security related exposures. The assessment is then used to develop a plan to prevent or counter the associated risks, threats, or vulnerabilities. Prior to looking at assessments, it's important to understand the differences between a RISK, VULNERABILITY, and THREAT. These words are frequently used interchangeably. In the realm of protection, they are NOT the same. Each is uniquely different. In the context of assessments, they can be done as three individual assessments or combined into one.

RISK

Percentage of likelihood of being targeted by a given attack, that attack being successful, and general exposure to a given threat.

VULNERABILITY

Security flaws or weaknesses in a system that allow an attack to be successful.

THREAT

Source and means of a particular type of attack.

In the context of protection, assessments are used to better understand your current situation or future position. This is often done from an adversarial perspective. It is an evaluation or estimation of your strengths, weaknesses, opportunities, threats, and capabilities. It is used to drive the decision-making process. It is also valuable as a tool to educate parties involved in the protective process. Common assessments are threat assessments, vulnerability assessments, and risk assessments. Each of these assessments have different objectives and often crossover. Some agencies and organizations combine these processes into a single assessment. These assessments are generally provided in a report-like format for consumption by select top-tier organizational members.

SWOT

A fast and commonly used process is a SWOT analysis. SWOT is an acronym for Strengths, Weaknesses, Opportunities, and Threats. This process is often used in business and can be performed quickly to better understand an organizations position. It is often done in a brainstorming format, but can also be outlined in a four-quadrant diagram. In the realm of protection, it is done from the perspective of the group and that of the adversary.

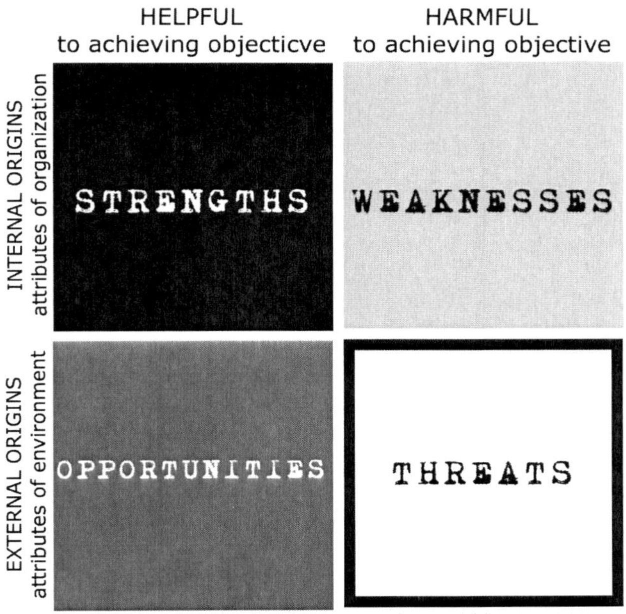

Examples:

STRENGTHS

- *Regional Experience*
- *Tactical Skills*
- *Language Skills*

WEAKNESSES

- *Understaffed*
- *Limited Finances*
- *No Local Licenses*

OPPORTUNITIES

- *Highly Regulated*
- *Asset Availability*
- *Local Contacts*

THREATS

- *Poor Infrastructure*
- *Frequent Violence*
- *Difficult to Blend In*

"Sound policy making requires ***recognition of reality****."*

-Secretary Pompeo

CARVER

A considerably more robust and complex system used in the protection world is the CARVER Methodology.

The acronym stands for:

Criticality
Accessibility
Recoverability
Vulnerability
Effect
Recognizability

CARVER is a threat and vulnerability assessment (TVA) system originally designed as a target analysis system to select the best possible military targets in state sponsored warfare. This system is said to have been designed by a US special operations officer, an Intelligence analyst, and a psychologist during the Vietnam era. In the protection field this system is reverse engineered and used to understand where and how an adversary may attack. The data is then used to strengthen or eliminate your weaknesses which is known as target hardening. This system has also been used to create matrix in the business world as well. The system creates perimeters for each of the six areas and assigns a point value to each potential outcome. The standard range is 1-5 with 1 being little to no value and 5 being of significant or high value. CARVER, like most processes in protection, can be used for people, places, and things.

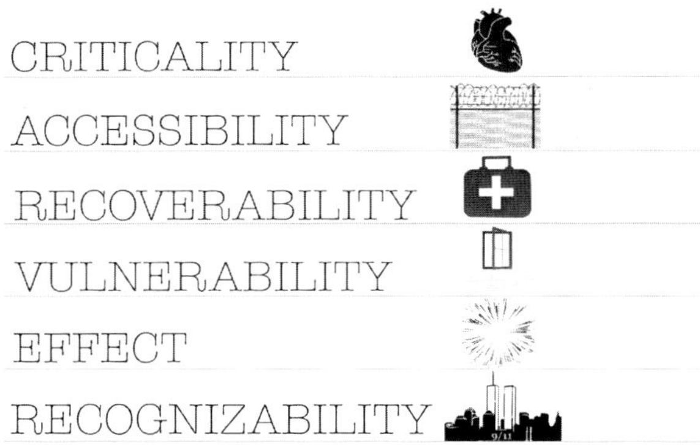

CARVER is usually rated on a **30-point scale**. It can be applied from the perspective of the attacker or defender. The highest score is a 30, which makes for a strong target choice. The lowest score is a 6, which is a difficult target to attack. In protection, we want to be the latter, as we are working to target harden.

EXAMPLE:

CRITICALITY *to human life*
Heart (5 points)
Arm (3 points)
Toe (1 point)

ACCESSABIILITY *to a business structure*
McDonalds (5 points)
Microsoft HQ (3 points)
White House (1 point)

RECOVERABILITY *of a human*
Human unprotected (5 points)
Human in body armor (3 points)
Human in MRAP vehicle (1 point)

VULNERABILITY *of a physical structure*
Glass (5 points)
Wood (3 points)
Steel (1 point)

EFFECT *of a physical attack on a human*
Bullet to Head (5 points)
Kick to Groin (3 points)
Slap to Face (1 point)

RECOGNIZABILITY *of an attack*
9/11 Attack (5 points)
Bill Gates Pie Attack (3 points)
School Yard Fight (1 point)

CARVER VALUE RATING SCALE

VALUE	C	A	R	V	E	R	VALUE
5	LOSS WOULD BE MISSION STOPPER	EASILY ACCESSIBLE AWAY FROM SECURITY	EXTREMELY DIFFICULT TO REPLACE. LONG DOWN TIME->1YR	DEFINATELY HAVE MEANS & EXPERTISE TO GO OFFENSIVE	FAVORABLE SOCIOLOGICAL IMPACT. OK ON INNOCENTS	EASILY RECOGNIZED BY ALL WITH NO CONFUSION	5
4	LOSS WOULD REDUCE MISSION PERFORMANCE CONSIDERABLY	EASILY ACCESSIBLE OUTSIDE	DIFFICULT TO REPLACE W/ LONG DOWN TIME -<1 YR	PROBABLY HAVE MEANS & EXPERTISE TO GO OFFENSIVE	FAVORABLE IMPACT. NO ADVERSE IMPACT ON INNOCENTS	EASILY RECOGNIZED BY MOST W/ LITTLE CONFUSION	4
3	LOSS WOULD REDUCE MISSION PERFORMANCE	ACCESSIBLE	CAN BE REPLACED IN RELATIVELY SHORT TIME-MONTHS	MAY HAVE THE MEANS & EXPERTISE TO GO OFFENSIVE	FAVORABLE IMPACT. SOME ADVERSE IMPACT ON INNOCENTS	RECOGNIZED WITH SOME TRAINING	3
2	LOSS MAY REDUCE MISSION PERFORMANCE	DIFFICULT TO GAIN ACCESS	EASILY REPLACCED IN A SHORT TIME-WEEKS	PROBABLY HAVE NO IMPACT	NO IMPACT. ADVERSE IMPACT ON INNOCENTS	HARD TO RECOGNIZE. CONFUSION PROBABLE	2
1	LOSS WOULD NOT AFFECT MISSION PERFORMANCE	VERY DIFFICULT TO GAIN ACCESS	EASILY REPLACED IN SHORT TIME-DAYS	DO NOT HAVE MUCH ATTACK CAPABILITIES	UNFAVORABLE IMPACT. ASSURED ADVERSE IMPACT ON INNOCENTS	EXTREMELY DIFFICULT TO RECOGNIZE W/O EXTENSIVE ORIENTATION	1

Human's don't hate violence. They hate violence taken out of context.

-Robert Sapolsky
(summary)

Chapter **EIGHTEEN**

HARD SKILLS

SHOOTING/FIREARMS

Most people have used firearms in some capacity or another in their lifetimes. Commonly they are used in a loose, familiarization setting like a local shooting range. In theory, it's a simple practice. You simply take the firearm (*a machine*) align the sights with a selected target, and press the trigger to the rear. The shot breaks and the round strikes the location where the sights where focused on the target. On the range this is rather easy to do. You know the weapon system, you know the target. You know the distance to the target. You are static. The target is static. Nobody is shooting back at you. Innocent people are not running in front of you. No hormonal heart rate, no tunnel vision, no auditory exclusion, no pooling of blood in your chest. So, you have a completely controlled environment that is artificial on all levels.

That's not to say that it doesn't have its place. Ideally a static/flat range is used for weapon familiarization and development of fundamental skills. This would be the *"crawl"* phase of firearms.

Once you have the basics down, it's time to walk, then run. Make sure induced STRESS is included in your training. This can be done by elevating your heart rate, limiting your senses, adding a physical penalty, or by competing against another human in the same task. Ensure that movement, yours and that of the target, is included in your training. Add elements of recognition into the timed training such as letters, numbers, colors, and shapes. Taking the time to properly identify these things is realistic, time-consuming, and stress-inducing. This makes the training both realistic and valuable. Low-light conditions, vehicle engagements, and multiple subjects *(shoot/no shoot)* should be integrated into training curriculums.

It makes sense for people who travel to learn the weapons systems used in the respective region of said travel. Even if you are not permitted to carry firearms in a given country, you may need to commandeer one from an aggressor in an extreme emergency. Imagine struggling, in a dire situation, to remove a weapon from a foreign adversary to find that you can't properly manipulate the weapon when his teammates arrive to greet you.

Basic examples of standard regional weapons listed below:

COUNTRY	COMMON PISTOL	COMMON RIFLE
US	Glock 17/19, Beretta M9	M4/AR-15, Barrett M107
MEXICO	HK P7 M13, Sig P226	H&K G3, HK PSG1
COLOMBIA	Cordova, IWI Jericho 941	Galil, Tavor
RUSSIA	MP443, Makarov	AK-47, Dragunov

Personal firearm selection will not be covered here. That subject is an entire book by itself. Selecting the right firearm is a lot like selecting the right clothing for a trip. You need to know the **environment** you will be operating in and the **activities** you will likely be participating in. These two factors should drive the decision-making process. Some other common factors may include hand size, carry style/location, and parts availability in regions where you will operate.

NOTE: *Consider taking a five-day firearms class at Roger's Range, Academi, or Gunsite.*

FIGHTING *(hand to hand, unarmed)*

Fighting to save your life or someone else's life is not the same as sport fighting. That said, you can't fight to the death in training, so gaps will always exist. The best way to close the gap is to find systems that are focused on applying the same techniques in training as in their real-world application. In other words, if a fighting style focuses on ripping out throats and gouging eyeballs, it is not likely realistic. You probably won't find a lot of training partners who will allow you to apply those techniques to them. This means these styles apply their techniques to the air or a bag. Those are simply not good substitutes for a human body. It's also an exceptionally unrealistic way of training for a high intensity battle where your adversary is fighting back with the intent **to kill you**. So, what does that leave? There are four core martial arts arts/sports that are appropriate for real, life threatening situations which happen under extreme stress. Each of the four, core combat sports are trained on humans who are fighting back and attempting to apply techniques. Each of them has proven effective over the course of many decades in extreme gladiatorial environments.

1. **Brazilian/Gracie Jiu Jitsu**
2. **Muay Thai/Kickboxing**
3. **Wrestling**
4. **Boxing**

There are many martial arts and sports that offer positive aspects. Most of them do offer fitness, flexibility, weight loss, discipline, confidence and belonging. The above picks are relative to what is tried and tested in the protection of humans. Again, **performance under stress must be part of the daily training regimen to be applicable in the realm of human protection.**

According to Tony Blauer, we need the ability to access *"protective aggression using primal gross motor movements fueled by fear and indignation."*

NOTE: *Consider dedicating three (3) hours per week to Brazilian Jiu Jitsu or Muay Thai/Kickboxing.*

DRIVING

Everyone drives, but not equally. Driving with the objective of departing point "A", integrating into local traffic movements on a controlled roadway, and eventually arriving at point "B" is a normal routine in modern life. This clearly changes for military personnel and law enforcement officers. They often pursue targets at high-speeds in *(sometimes)* less than ideal environmental conditions. Keep in mind that PROTECTIVE DRIVING is the opposite. The protectee is often being pursued and the objective is often to lose or separate from the pursing individual or group. This means a major change in mindset is required. It also means that tactics need to change relative to the situation. Another area that differs in protective driving is convoys or motorcades. This means all drivers in the motorcade need to know the same playbook and be able to implement it on demand.

NOTE: *Consider taking a three-day* ***Security Driver Course*** *at* ***BSR****. Before you go, have a look at an NIJ armor performance standards chart.*

SURVEILLANCE DETECTION (BASIC)

Part of human understanding in protection encompasses what is occurring in your environment. Identifying a baseline and detecting an anomaly is a core process that is continually repeated. One area that needs to be considered and assessed is surveillance. There are several types of surveillance, but the main types include passive and active, static and mobile. If someone is conducting surveillance, their response to your detection varies based on their training and experience.

Identification of surveillance includes looking for unusual behaviors or activities, people running away when you make eye contact, adults hiding out of context, or sudden unreasonable reactions.

The acronym **TEDD** offers a quick way to assess your momentary situation when you may be under surveillance or being followed. This system is common in the government and private sector.

Time - *Repeated sightings over time*
Environment - *In different, unrelated realms*
Distance - *Over distant locations*
Demeanor – *If they move when you move. If they make sporadic foot or vehicle movements. If they use hand and arm signals.*

MEDICAL CONSIDERATIONS (BASIC)

Medical skills vary considerably in protective service operations. Knowledge levels start at basic first aid and CPR and extend upward to trauma surgeons. Most people in protection don't need to learn how to clean a small cut and apply a bandage. They need to focus their attention on significant, live threatening issues. In this particular case, the focus is on pre-hospital trauma. One of the positive sides of continuous military combat operations across the globe is the understanding and development of trauma-based emergency care. A system called Tactical Combat Casualty Care (TCCC) incorporated that knowledge into one place. Their system takes into consideration the hostile environment, the severity of the medical issue, evacuation options, field care, transfer of care, and more. They also have process flow charts referred to as "algorithms" that assist the user in making critical decisions. Some of the areas addressed are massive hemorrhaging, airway management, respiration/breathing, traumatic brain injury, hypothermia, burns, splints, and wound care. A number of systems and courses have sprung up in recent years attempting to meet the standard of training TCCC offers. To date, TCCC remains the gold standard.

NOTE: If training at the TCCC level is NOT accessible, consider finding training in some of the core areas of trauma care below.

Tourniquets - *function & application*
Hemostatic agents – *function & application*
Israeli bandages - *function & application*
Chest seal - *function & application*
Immobilization splints - *function & application*

RESCUE ESSENTIALS is a great supply source in NC.

PHYSICAL STRUCTURING

Setting up a physical security structure is part and parcel of protecting someone or something. The most frequent method deployed by protection professionals for physical security is known as the ***Layered Security Approach***. It can be called many different things, but the end result is the same, **concentric rings** of protection are established. One system is called the **DDDRM** process.

DETER

Deter the adversary from ever attempting the attack. This is the outer layer. It could be a fence, concertina wire, dogs, cameras, roving patrols, etc.

DETECT

Detect when the adversary breaches the structure. This can be achieved with sensors, cameras, guards, dogs, or other methods.

DELAY

Delay the adversary once they have breached and been detected. This delay provides you with more time to respond. They can be achieved by creating a maze-like system and having access points on opposite sides of the layers.

RESPOND

Physically respond to the adversary's breach. This is typically a human response. This element is often done by a quick reaction force or QRF.

MITIGATE

This portion is the management of the end results of damage by the adversary or the impact to the adversary by the QRF. This can become an area for legal intervention.

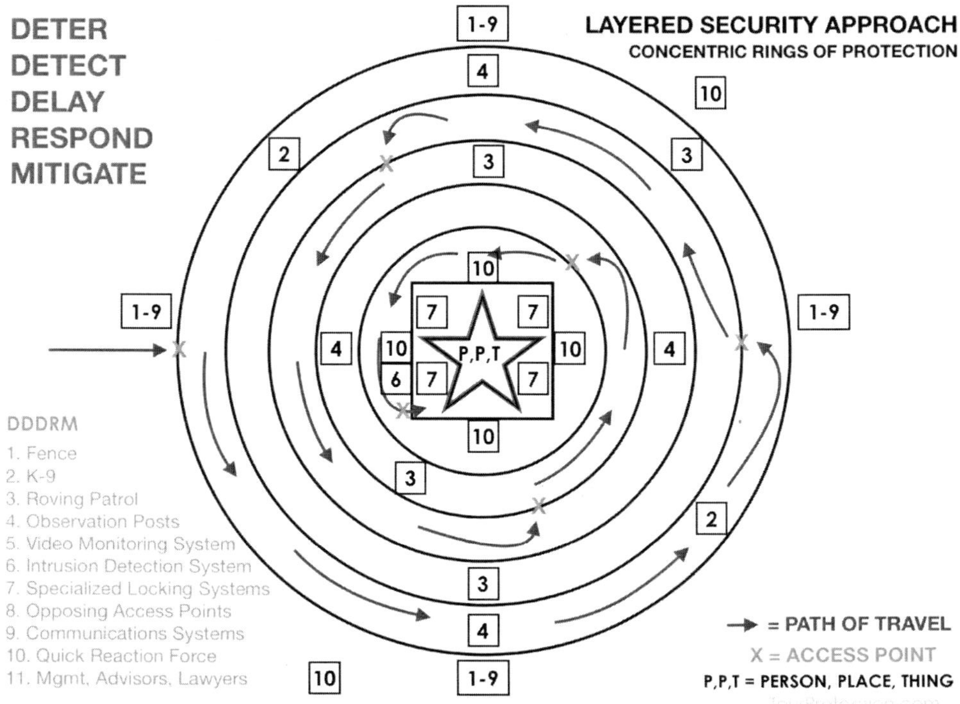
LAYERED SECURITY APPROACH
CONCENTRIC RINGS OF PROTECTION
DETER
DETECT
DELAY
RESPOND
MITIGATE
DDDRM
1. Fence
2. K-9
3. Roving Patrol
4. Observation Posts
5. Video Monitoring System
6. Intrusion Detection System
7. Specialized Locking Systems
8. Opposing Access Points
9. Communications Systems
10. Quick Reaction Force
11. Mgmt, Advisors, Lawyers
1-9
4
2
3
10
7
P,P,T
6
= PATH OF TRAVEL
X = ACCESS POINT
P,P,T = PERSON, PLACE, THING

Assessments of physical structuring require adversarial thinking. Some areas include ZOP, SOF, and KOCOA, and Mx7, all of which are defined in this chapter and the next.

Zone of Predictability (ZOP)

A zone of predictability is an area or space where there is a high level or predictability for a specific action by a specific individual or group. Attackers need to identify these zones during pre-attack surveillance. Protectors need to identify these zones to ensure a heightened sense of awareness while inside of these zones.

Standoff Footprint (SOF)

A standoff footprint is a location from which an adversary can initiate an attack on a given target. The location needs to have visual access to the target. The individual initiating the attack does not have to be a physical attacker. They may simply be operating in a communications capacity and deploy the attackers following visual confirmation of location. Again, protectors need to identify these locations to ensure a heightened sense of awareness while in close proximity to potential SOFs.

TERRAIN ANALYSIS

Terrain analysis is the collection, analysis, evaluation and interpretation of the geographic information on natural and manmade features of the terrain, combined with other relevant factors, to predict the effect of the terrain on an operation or activity. It is conducted by many people for many different reasons. It's done by engineers, geologists, ecologists, archaeologists, landscape designers, and city planners to name a few. In protection, the analysis is most similar to that of the military.

KOCOA

KOCOA is an acronym used by a number of military organizations for basic terrain assessments. It is of equal value for individuals functioning in a protective capacity.

Key Terrain
Observation & Fields of Fire
Cover & Concealment
Obstacles & Obstructions
Avenues of Approach & Escape

KEY TERRAIN. What terrain in your area will likely afford you success in the event of an attack? What terrain in your area will likely create your demise in the event of an attack?

OBSERVATION & FIELDS OF FIRE. What positions offer the ability to quickly and safely see activity in the area? What positions afford the ability to quickly and accurately engage threats with firearms?

COVER & CONCEALMENT. Which areas offer ballistic protection from incoming small arms attacks *(cover)*? Which areas offer the ability to hide from adversaries *(concealment)*?

OBSTACLES & OBSTRUCTIONS. What land features or artificial elements present problems or restrict movements and actions for both you and the adversary?

AVENUES OF APPOACH & ESCAPE. Which direction(s) could your adversary approach from? Which direction(s) would likely offer the ability to escape from an adversary's approach or attack?

*What areas are **Key Terrain**?*
*What areas offer **Observation & Fields of Fire**?*
*What areas offer **Cover & Concealment**?*
*What areas present **Obstacles & Obstructions**?*
*What areas offer **Avenues of Approach & Escape**?*

Below you will find several common examples for KOCOA. These are general examples and will change with each environment and landscape.

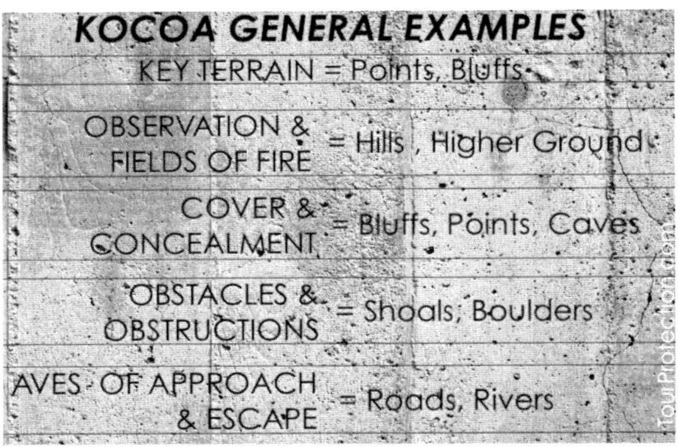

"We don't rise to the level of our expectations. We fall to the level of our training."

-Archilochus

SYSTEM/PROCCESS WEAKNESSES

All systems and processes have weaknesses or vulnerabilities. These weaknesses allow for potential exploitations to occur. There are seven (7) exploitable weaknesses commonly found in most organizations/businesses. These are referred to as ***Mx7****. In protection, these seven areas require special attention as well as hardening measures.*

Machines/Tool

Weapons, Vehicles, Computers, Phones, etc.

Methods

Tactics, Techniques, Procedures, Field Craft, etc.

Measurements

Time, Distance, Units, etc.

Materials

Information, Data, Propaganda, etc.

Manpower

Contacts, Personnel, Officials, Threat Actors, etc.

Mother Nature

Storms, Tornados, Earthquakes, Floods, etc.

Money

Sources, Movement In/Out, Duration, Location, etc.

"An ounce of prevention is worth a pound of cure."

-Ben Franklin

Chapter **NINETEEN**

ADVANCING

"*Advancing*" is the choreography of protection. It is essentially information collection and distribution, typically performed in advance of the protectee's arrival to a key location. It can be done via email, phone, and location walk-throughs or, *preferably*, **all three combined**. In this section we will look at hotels, airports, restaurants, and events. Each location will have crossover points and unique areas of attention or concern.

HOTELS

During business trips, personal vacations, or other travel, many people will spend a considerable amount of time in a hotel. In the business world, it is common to hold meetings in hotels as well. Hotels are the traveler's home away from home. Almost all daily movements will originate and terminate at the hotel. This produces two points which should be considered.

The first area of concern is the ease of which one can be sucked into the comfort of a resting place (home). However, unlike your home, a high-volume of unknown people are coming in and out daily. Privacy is only a myth inside a hotel, even inside your room. Most all staff members have master keys. They also possess access to your details on file in the system such as your full name, home address, phone number, credit card number, and possibly your passport if you are overseas.

The second area of concern is that your function or activities can be revealed or exposed to a number of other guests or people operating inside the hotel. As noted above, most movements originate and terminate at this location. This creates a zone of predictability (ZOP). Anyone can easily follow your movements (tail) when you depart. Additionally, it's not difficult to piggyback into an elevator and follow a given person to their hotel room.

Other areas of awareness include unsecure phone and internet systems which can easily be compromised.

There are a number of additional considerations that must be made if staying with a client. Here are some key areas to look at:

LOCATION. The location of the hotel can play a key role in safety. Where is the hotel located in town? What is the history of the area? What activities are common there? Is it situated in an urban or rural area? How far is it from the police, fire, ambulance, hospital, or embassy?

STRUCTURE. The hotel design can hurt you or help you. What is the building made of *(construction materials)*? How many floors does it have? How easy or difficult is it to identify?

CONTACTS. This is your conduit throughout the stay. Who is your primary contact at the hotel? Do you have additional contacts? GM, AGM, Front Desk Manager, Security Manager, etc.?

SECURITY DEPARTMENT. This may be a "force multiplier" or completely useless. Does the hotel have a dedicated security department? If so, who manages it? How many staff do they have? What are their rotation shifts? Do they have a duty phone?

VIDEO MONITORING SYSTEM (VMS). Does the hotel have a video monitoring system? If so, is that system IP based or hard wired? Who has access? What types of cameras *(fixed, PTZ, etc.)*? What are the camera locations? Can you obtain video copies if there is an issue? How long is the data stored?

ACCESS CONTROL SYSTEMS. Are searches conducted coming into the hotel *(guests, staff, etc.)*? Is there a system to allow only guests to get past the lobby? Are the elevators key card controlled? What type of locking systems are used for room doors?

ELEVATORS. How many public and service elevator banks does the hotel have? What are their locations? Are they key card controlled? Do they have independent motion/lock out capability?

LOADING DOCKS. Loading docks serve as a point for deliveries and pick-ups. They also serve as an alternative exit.

ROOF. The roof can be a useful location to access in the event of an emergency. Is the roof accessible? What is the layout? What is the ground material *(concrete, shingles, gravel, etc.)*? Could a helicopter land on the roof?

PARKING. Parking structures, garages, or lots typically exit at hotels. Once again, the layout and design can either assist you or potentially harm you. Is the parking structure below ground? Is there an elevator that connects to room floors? Is there a parking lot? What is the lighting level at night? What is the distance to the front door of hotel?

MEDICAL. In a critical incident, time is essential. Access to a hospital may be limited or unrealistic. You will be limited to what gear you brought and what the hotel has on-site. Does the hotel have a trauma kit? Do they have hemostatic agents? Do they have AEDs? If so, what are the locations? Do they have a backboard? Do they have a wheelchair? Does the hotel have an in-house doctor?

FLOOR PLANS. Prior to 9/11, hotel floor plans were easy to obtain. As time moves on, they have become easier to access. If a hotel will not provide you floor plans, you will likely find one on the back of the door inside the room. Most room floors will be identical.

FLOORS. Learning the hotel is an element that can pay dividends in an emergency. The first step is getting a list of what is located on each floor. Where is the lobby located? Restaurant? Bars? Gym? Spa? Medical?

ROOMING LIST. If you are staying with a group, or even another person, you should be aware of their location inside the hotel.

PHOTOS/SKETCHES. A picture is worth a thousand words. Take photos when possible. When photos are not possible, sketch it out. This will ingrain the layout into your head and will also allow you to share with others as needed.

OTHER VIPS IN HOTEL. In the realm of protection, this is called deconfliction. It's important to know who is around you and what attention they will draw to the property. Avoid moving into or out of the hotel when VIPs are coming or going. They may draw chaos and increase the risk of a critical event. Staying away from this could prevent you from being in the wrong place at the wrong time.

OTHER EVENTS IN HOTEL. Know what companies or events are occurring inside the hotel during your stay.

POLICE, FIRE, HOSPITALS, EMBASSIES. Once again, it is paramount to protection to know where the key city locations are in relation to the hotel. We suggest you map out police stations, fire stations, hospitals, and embassies prior to your arrival.

1st FLR

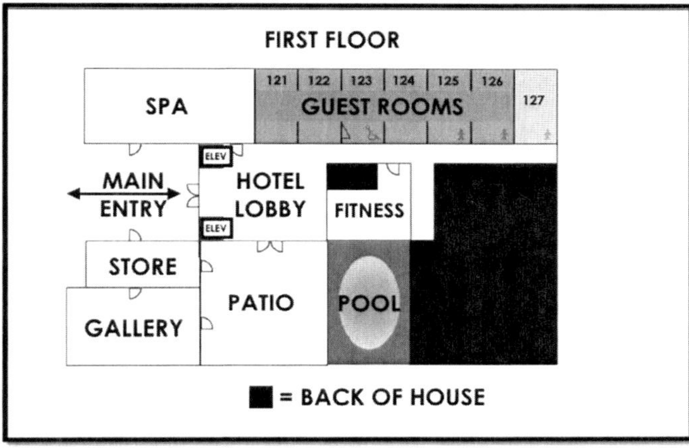

2nd FLR

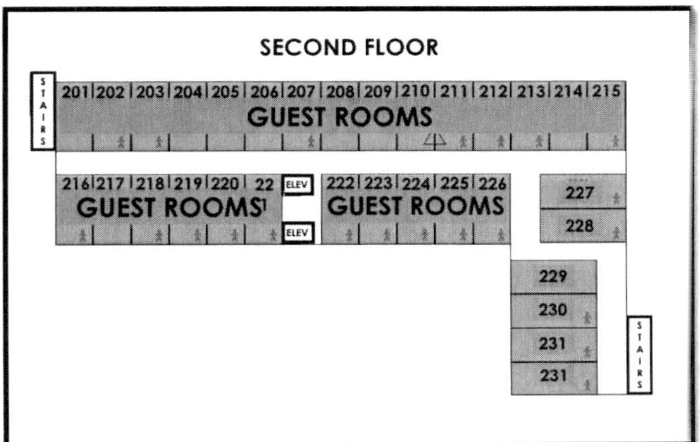

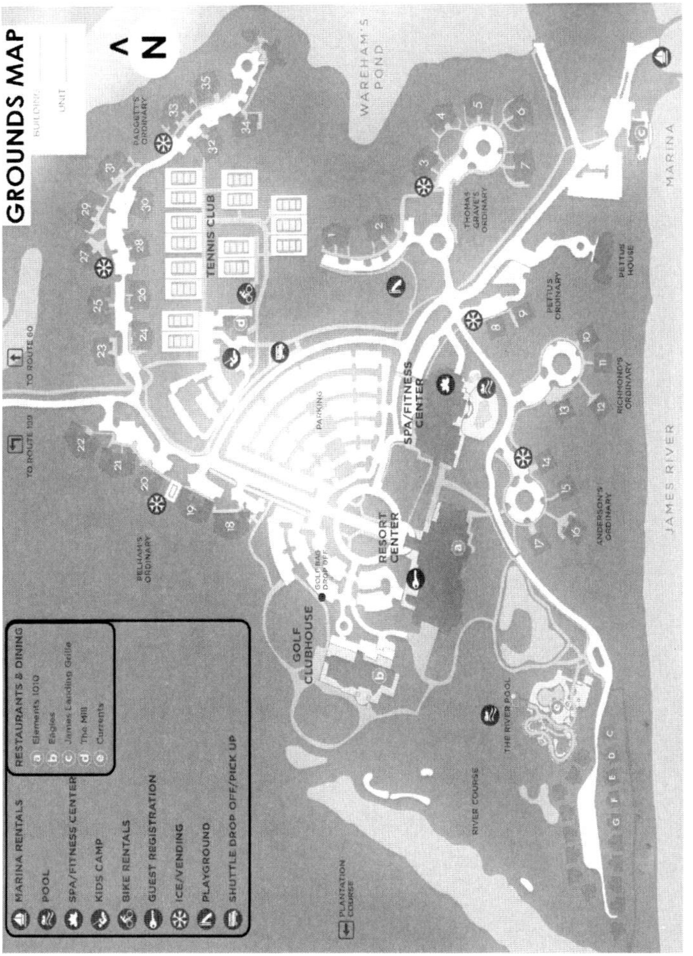

GROUNDS MAP
BUILDING
UNIT
N
TO ROUTE 60
TO ROUTE 199
WAREHAM'S POND
PADGETT'S ORDINARY
TENNIS CLUB
THOMAS GRAVE'S ORDINARY
PETTUS ORDINARY
PETTUS HOUSE
MARINA
PELHAM'S ORDINARY
PARKING
SPA/FITNESS CENTER
RESORT CENTER
RICHMOND'S ORDINARY
ANDERSON'S ORDINARY
JAMES RIVER
GOLF BAG DROP OFF
GOLF CLUBHOUSE
THE RIVER POOL
RIVER COURSE
PLANTATION COURSE
MARINA RENTALS
POOL
SPA/FITNESS CENTER
KIDS CAMP
BIKE RENTALS
GUEST REGISTRATION
ICE/VENDING
PLAYGROUND
SHUTTLE DROP OFF/PICK UP
RESTAURANTS & DINING
a Elements 1010
b Eagles
c James Landing Grille
d The Mill
e Currents

AIRPORTS

Airports are another common place that is frequently utilized by celebrities, executives, and dignitaries. They are also frequented by families and friends who travel for work or vacation. The most common types of airports are commercial and private, or fixed base operator (FBO). Most people traveling for vacation or business use commercial airports and flights due to the affordable costs and number of available flight options. However, many celebrities, executives, and dignitaries prefer to fly private and have little concern with the cost implications. Some dignitaries will have access to a third airport type located on military bases. This is the least common means of travel.

LOCATION. Commercial airports and private airports, often known as FBOs, can be located on the same property or in separate locations in different parts of the city. It is strongly suggested that location confirmation is done well in advance of arrival. The easiest way to do that is with a three-letter airport code. This will automatically assist you with finding the correct airport.

FLIGHT DETAILS. It seems like a no-brainer to say that flight details are required in advance of arrival, but frequently protectees will arrive without essential details. This means it is the protector's job to have the information in advance of arrival. Airlines, flight number, departure and arrival times are the basics. When flying private, it's essential that you have the aircraft type, **tail number**, and/or call sign.

AIRPORT POINTS OF CONTACT. Having a primary POC for the airport is crucial. Additionally, having specific contacts such as the Operations Manager and Security Manager often prove invaluable.

ACCESS. There are a few considerations when it comes to accessing controlled areas of an airport. The first is vehicle access to the tarmac. Ideally, in private airports, the protectee would be driven through a security check point and taken to the door of the aircraft on the flight line. This has become hit and miss since 9/11. Smaller cities generally permit this access while larger ones commonly restrict it. In commercial airports a holding area, or VIP lounge is preferred for the protectee. This keeps them in a more controlled space, often with people of a similar status. While status can be bought, it is still preferable over waiting at a general gate.

MISC. DETAILS. Who will greet the protectee upon arrival to the aircraft? Will there be local handlers who will facilitate luggage going into the rear of the aircraft? Will there be press on-site? If so, which outlets and where will they be located? Does the protectee intend to speak? What is the timeline and how does this impact the "wheels up" time?

INTERNATIONAL. Make sure you and your protectee have your passports on your person. When possible, get a customs and immigration point of contact prior to arrival. Cell phone numbers always work best. Know if they will board the plane or if the protectee will need to come off and pass through a standard process.

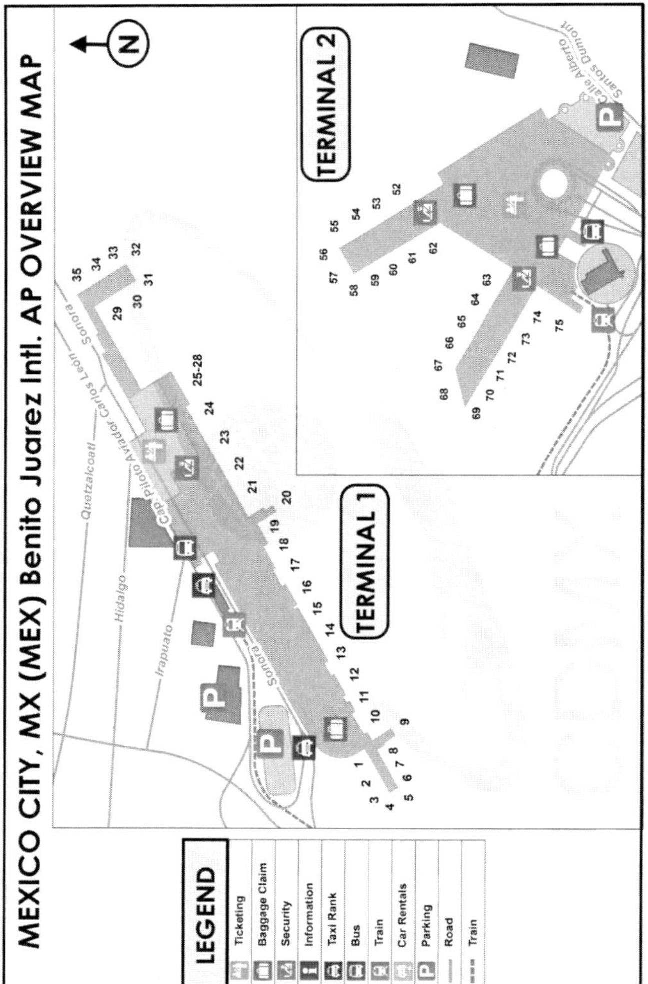
MEXICO CITY, MX (MEX) Benito Juarez Intl. AP OVERVIEW MAP
N
TERMINAL 1
TERMINAL 2
Sonora
Hidalgo
Irapuato
Quetzalcoatl
Cap. Piloto Aviador Carlos Leon
Calle Alberto Santos Dumont
LEGEND
Ticketing
Baggage Claim
Security
Information
Taxi Rank
Bus
Train
Car Rentals
Parking
Road
Train

RESTAURANTS

Restaurants are a cornerstone in any area of entertainment, government, or business operations. They host meetings and celebrations for clients and families alike. Anyone performing protective services will spend a great deal of time in restaurants.

Some of the key areas of consideration for advancing are listed below.

-Owner/General Manager Contact
-Verbal Privacy Agreement
-Vehicle Drop Off/Pick Up Locations
-Entry/Exit Options
-Dining Room Layout
-Room Sight Lines
-Private Dining Options
-Windows & Walls
-Client Seating Selection
-Protection Detail Seating
-Bathroom Location
-Safe Haven/Temp Hold Options
-Server Selection & Meeting
-Menu Copy
-Closest Trauma 1 Hospital
-Closest Police Station

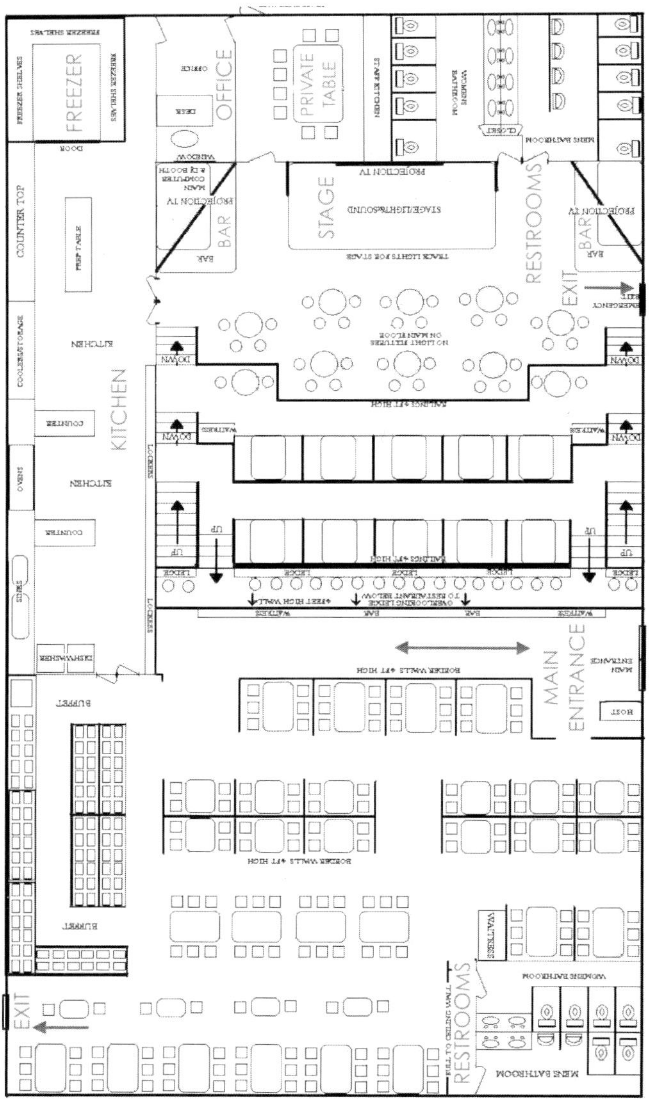
FREEZER
OFFICE
PRIVATE TABLE
COUNTER TOP
BAR
STAGE
RESTROOMS
BAR
EXIT
KITCHEN
MAIN ENTRANCE
EXIT
RESTROOMS

VENUES & EVENTS

Events are common among dignitaries, celebrities, corporations, and charities. They are also common with every day families. Events can range from a small group of 40-50 people to hundreds of thousands of people. The larger the event, the more complex they become. This is in part because of the number of variables that exist regarding the scale. It is also because events that are large can sway human emotion, which can equate to actions or behaviors that are not always positive. They also allow individuals (threat actors) to blend into innocent crowds. This cover affords them freedom of movement and cover for action.

Planning for these events is time consuming. It starts with establishing contacts. These contacts are typically section or department heads for a given event. Once established, the most common, modern approach is to open a channel of communication. This is done via phone or email. Questions and answers are exchanged in a number of formats, but for protective service operations, they are typically done via written advance. This advance may start with questions and evolve into a list of directives or requests for the event managers. These start the official process and also establish documentation.

There are many factors to consider and modify based on the client's specific needs. Some of the core areas related to events are listed below. This could produce a 500-page book alone.

-Distribution of Security Materials
-Appropriate Related Client History
-Related Event History
-Anticipated Crowd Demographic
-Religious/Political Content
-Communications Plan
-Deconfliction
-Event Stop Procedures
-Terrorism/Active Shooter/Bomb Threat/ Response protocols
-Client Holding & Extraction Options
-Patron Emergency Egress Options
-Plan Development
-Threat Assessment
-Vulnerability Assessment
-Risk Assessment & Matrix

-Credential Systems
-Restricted Items
-Signage Required
-Audio Loops
-Video Loops
-Employee Searches
-Patron Searches
-Police, Medics, Fire
-Barriers
-Pyrotechnics/Strobes/Audience Plants
-Concessions
-Patron Hydration
-Crowd Migration Plan
-Police Escort
-Staffing/Deployment Maps
-Post Orders
*-Post Event **A**fter **A**ction **R**eport/Review*

Following the distribution of this written advance document, the protective service team would have an advance phone call to ensure a live and dynamic exchange between involved parties.

The final advance is physical. In protective service operations, this is generally done in the format of a "walk-through" at the event site hours, days, weeks, or months in advance.

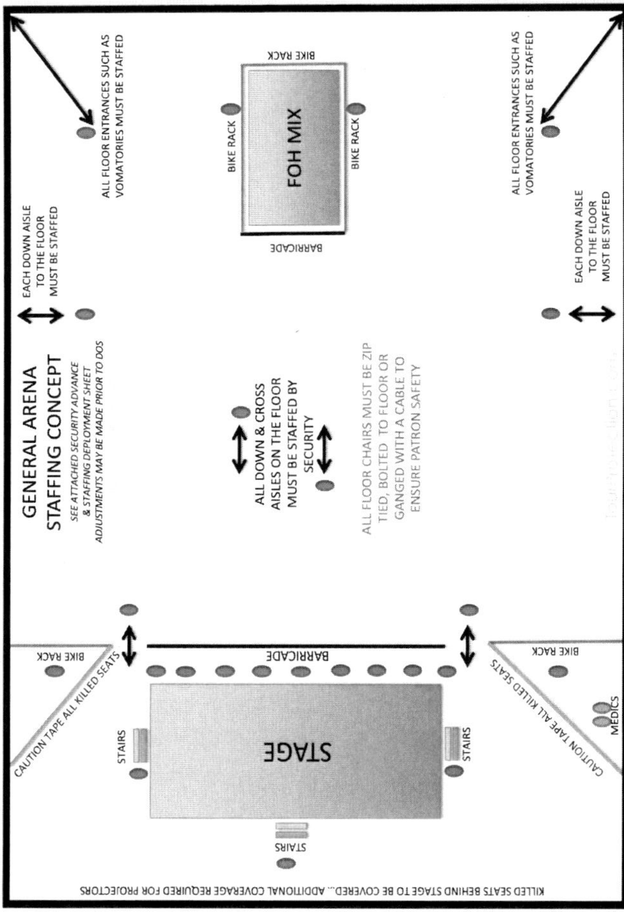
GENERAL ARENA
STAFFING CONCEPT
SEE ATTACHED SECURITY ADVANCE
& STAFFING DEPLOYMENT SHEET
ADJUSTMENTS MAY BE MADE PRIOR TO DOS
EACH DOWN AISLE
TO THE FLOOR
MUST BE STAFFED
ALL FLOOR ENTRANCES SUCH AS
VOMATORIES MUST BE STAFFED
BIKE RACK
FOH MIX
BIKE RACK
BIKE RACK
BARRICADE
ALL FLOOR ENTRANCES SUCH AS
VOMATORIES MUST BE STAFFED
EACH DOWN AISLE
TO THE FLOOR
MUST BE STAFFED
ALL DOWN & CROSS
AISLES ON THE FLOOR
MUST BE STAFFED BY
SECURITY
ALL FLOOR CHAIRS MUST BE ZIP
TIED, BOLTED TO FLOOR OR
GANGED WITH A CABLE TO
ENSURE PATRON SAFETY
BIKE RACK
CAUTION TAPE ALL KILLED SEATS
BARRICADE
STAIRS
STAGE
STAIRS
STAIRS
BIKE RACK
CAUTION TAPE ALL KILLED SEATS
MEDICS
KILLED SEATS BEHIND STAGE TO BE COVERED... ADDITIONAL COVERAGE REQUIRED FOR PROJECTORS

General Staffing Dot Map

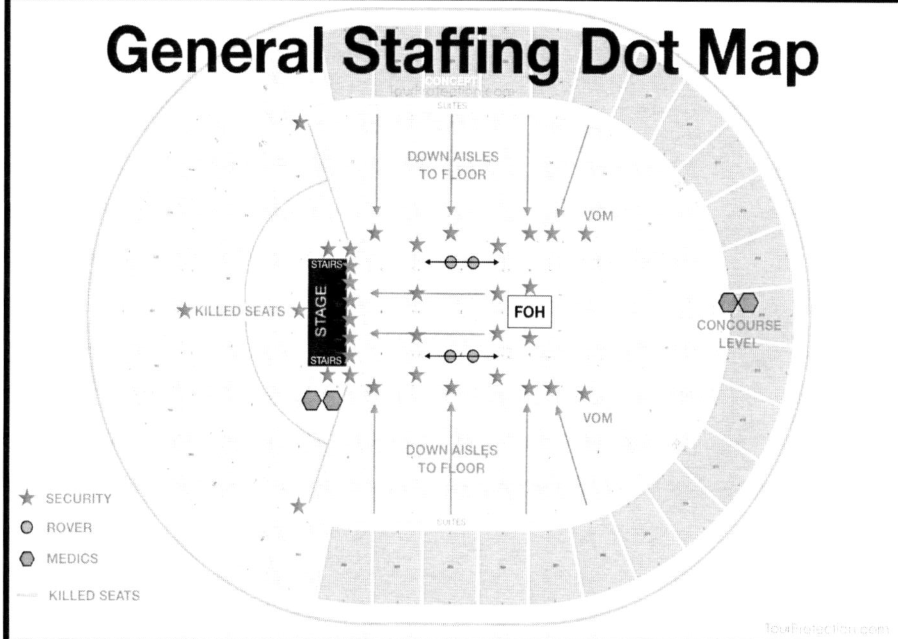

VEHICLES *(General Concept)*

Two types of vehicles are used from a protective perspective. The preferred vehicle type is a premium, high-center of gravity (SUV) that is capable of jumping large curbs or pushing through obstructions in the roadway. The secondary option is a premium, sedan with a high-output motor (FAST) capable of creating space between the protectee and potential problems. Each of the two vehicle types require a minimum of four (4) doors and must have seating for one (1) security guard (passenger front) and protectee(s) (rear). There needs to be significant storage space in the back of the vehicle or trunk. It's suggested that the vendor provide photos of the interior and exterior of the vehicle for approval and confirmation prior to arrival. Once vehicles are approved, vehicle IDs, armor certs, and license plates should be provided. The follow vehicle (security vehicle) needs to be equal in performance capability and size. It must be four (4) doors and have room for at least three (3) passengers with the ability to function as a back-up vehicle for the protectee. Again, advance photos of the interior and exterior of the vehicle should be sent by vendor for approval and confirmation prior to arrival. Again, once vehicles are approved, vehicle IDs, armor certs, and license plates should be provided.

DRIVERS *(General Concept)*

We require that all drivers are security trained. Ideally, they will maintain certification with a recognized protective driver association or have a military, law enforcement, or embassy driver history. All drivers should be expected to know the local areas and associated roads/streets. We require the driver's name, cell phone number, license and photo prior to arrival. We require the driver to arrive daily with a full tank of gasoline, manufacturer's recommended tire pressure, clean inside and out, with all doors/windows/locks functioning appropriately. All vehicles should have spare tires and associated tools. These areas should be checked frequently throughout the trip. The radio must remain off unless otherwise directed by the protectee. AT NO TIME SHOULD THE DRIVER USE HIS OR HER CELLULAR PHONE WHILE OPERATING THE VEHICLE. Drivers who divulge any information about the protectee's location, activities, or personal/business associations will be immediately dismissed from their duties and will not be paid for that day. This includes social media postings of ANY type. Drivers should not eat, drink, or smoke in the protectee vehicles. At NO time during the work day will the driver be permitted to use alcohol or drugs. Vehicles should arrive clean on the interior and exterior. Drivers' attire should be business casual unless otherwise directed. Ensure that the driver is afforded ample time to eat, drink, and use the restroom.

FORMATIONS *(General Concept)*

Vehicle formations vary from basic to complex. They are often based on available assets, protective strategies, and local law enforcement support. The most common formations include two to three vehicles. Lead, Limo (Principal), and Follow (Chase) are the most typical positions. Motorcades (convoys) are designed to function as a single protective unit but with different, individual tasks in the event of an attack on the motorcade. Vehicle roles may be screening, shielding, package (principal transfer), pushing through contact, counter-assault (CAT) or other specific function. Driving as a team requires extensive training and experience, as each driver must have a strong understanding of what other drivers will do in each circumstance.

Some of the **most basic** movement techniques include:

1) Offset Driving
2) Lane Clearance
3) Rear End Protection/Turns
4) Screening/Shielding
5) Disabled V/Package Transfers
6) Contact Techniques
7) Driver Down Techniques

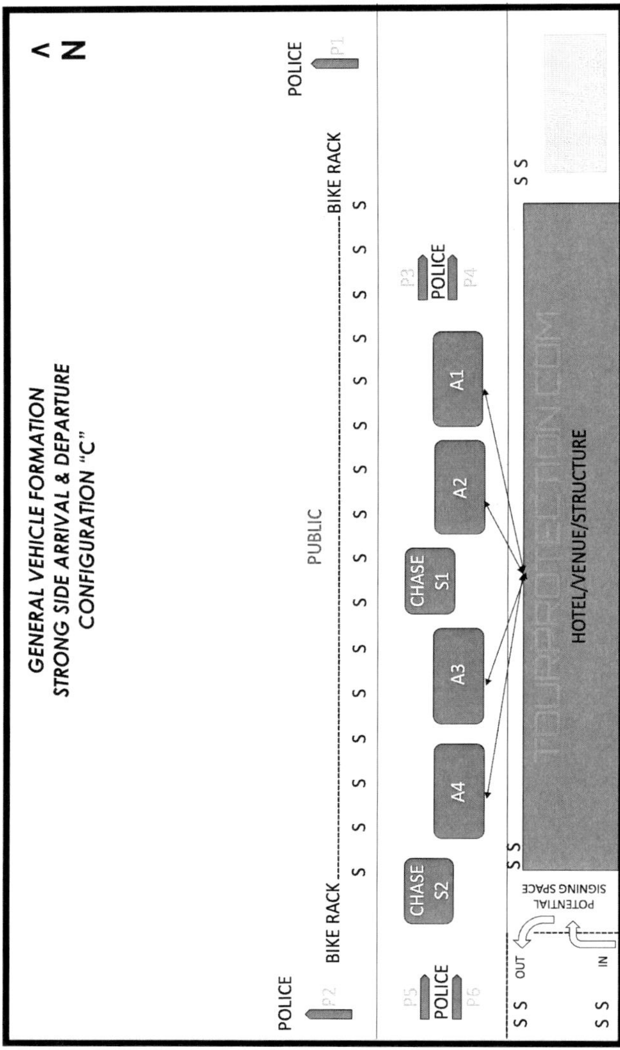
GENERAL VEHICLE FORMATION
STRONG SIDE ARRIVAL & DEPARTURE
CONFIGURATION "C"
N
POLICE
P1
BIKE RACK
PUBLIC
BIKE RACK
POLICE
P2
P3
POLICE
P4
A1
A2
CHASE S1
A3
A4
CHASE S2
P5
POLICE
P6
POTENTIAL SIGNING SPACE
OUT
IN
HOTEL/VENUE/STRUCTURE

VEHICLE/DRIVER HASTY CHECK LIST

VEHICLE SELECTION & INSPECTION
(When possible)

Determine the best vehicle option for the environments and the protectee's personal needs. This typically breaks down into two categories.

1) **High Center of Gravity** (SUV)
   The ability to push through obstacles and jump curbs.

2) **High Output Motor** (Fast Luxury)
   The ability to create space between the protectee and the threat.

VEHICLE HASTY LIST

- *Interior/Exterior Sweep*
- *Full Tank of Gas*
- *Tire PSI*
- *Windows & Locks Function (DEA)*
- *Windows Up, Doors Locked*
- *Spare Tire*
- *Radio Off*
- *A/C Set*
- *Car Clean*

DRIVER SELECTION & BRIEF

(When possible)

Learn the potential driver's history in:
-Evasive and protective driver training, along with experience
-Security training, along with experience
-High-risk environment exposure (Mil/LE) along with overseas experience

Ensure that a detailed discussion of the routes selected and emergency protocols or *Tactics, Techniques, & Procedures* (TTPs) are had prior to departure.

Discuss and confirm all **times** and **distances** *to and from* all potential locations to be visited by the protectee.

HASTY DRIVER BRIEF

- *No Chitchat*
- *No Cell Phones*
- *No Social Media*
- *Driver Remains in Vehicle*
- *Common Vehicle Formations*
- *Use of Placards & Placement*
- *Vehicle Formation & Positioning*
- *Car Door Stays Open Until Principal is Inside Building*

"Everywhere is nowhere. When a person spends all his time in foreign travel, he ends by having many acquaintances, but no friends."

-Seneca

Chapter TWENTY

FOREIGN TRAVEL

Travel equates to a significant environmental change. In protection, this requires a shift in both mindset and tactics. Business and personal travel both present risk exposures from a protective standpoint. While domestic travel presents potential risks, foreign travel presents the greatest and most serious risk exposures to the traveler. Taking some basic precautions before and during travel will assist in offsetting many of the traveler's risk exposures.

- *Learn about the local culture and history*
- *Learn basic phrases in the local dialect*
- *Always appear to be with others*
- *Dress and move like a local*
- *Blend in when possible*
- *Mirror behavior/mannerisms*
- *Avoid tourist traps/hangouts*
- *Have contact info for hospitals*
- *Have safe haven options*
- *Have a plan*

DOS

Use the U.S. Department of State (DoS) as your personal asset. If you're American, that's what they are there for. For other nationalities, use your country's equivalent. The DoS website offers a quick look at security considerations for Americans traveling to foreign countries. This is likely to be the single best contact when it comes to gathering country-specific information.

RSO

Each U.S. Embassy has a head of security known as the Regional Security Officer, more commonly referred to as the RSO. This individual is tasked with overseeing all security and safety related issues in a host country, including the protection of the Ambassador.

The RSO is an excellent asset for professional security providers. If you are protecting dignitaries, high-profile executives or celebrities, you will likely need to speak with the RSO. While the

RSO is an important person with valuable information *(much of which can't be shared)*, you should *NOT* consider contacting them unless it is essential to your trip.

OSAC

OSAC stands for the Overseas Security Advisory Council. It is a partnership, exchange, and communication platform between the security arm of the DOS (DS) and US private sector businesses. The OSAC "*Council*" comprises 31 private sector and 3 public sector member organizations that represent specific industries or agencies operating abroad. The member organizations designate representatives to serve on the Overseas Security Advisory Council to provide direction and guidance to develop programs that most benefit the U.S. private sector overseas. A primary goal of OSAC is to develop an effective security communication network, consequently, OSAC invited all U.S. businesses, academia, faith-based groups, and non-governmental organizations to become constituents.

There is no cost involved with OSAC constituency. OSAC provides a number of frequently updated, open source documents that contain valuable information to those working in protective service operations.

FOREIGN COUNTRY DATA (BASIC)

- *Language spoken*
- *Currency used*
- *Common criminal activity*
- *Criminally controlled areas*
- *Embassy location*
- *Safe haven options*
- *High quality hospitals*
- *Climate & weather*
- *Terrain & altitude*
- *Water & food safety*
- *Local activities & events*
- *Review CDC advisory*
- *Kidnap & Ransom (K&R) insurance*

MAPPING - FOREIGN TRAVEL (BASIC)

- Airports
- Hotels
- Venues
- Hospitals
- Safe Havens
- Embassy/Consulates
- Evacuation points *(air, land, sea)*

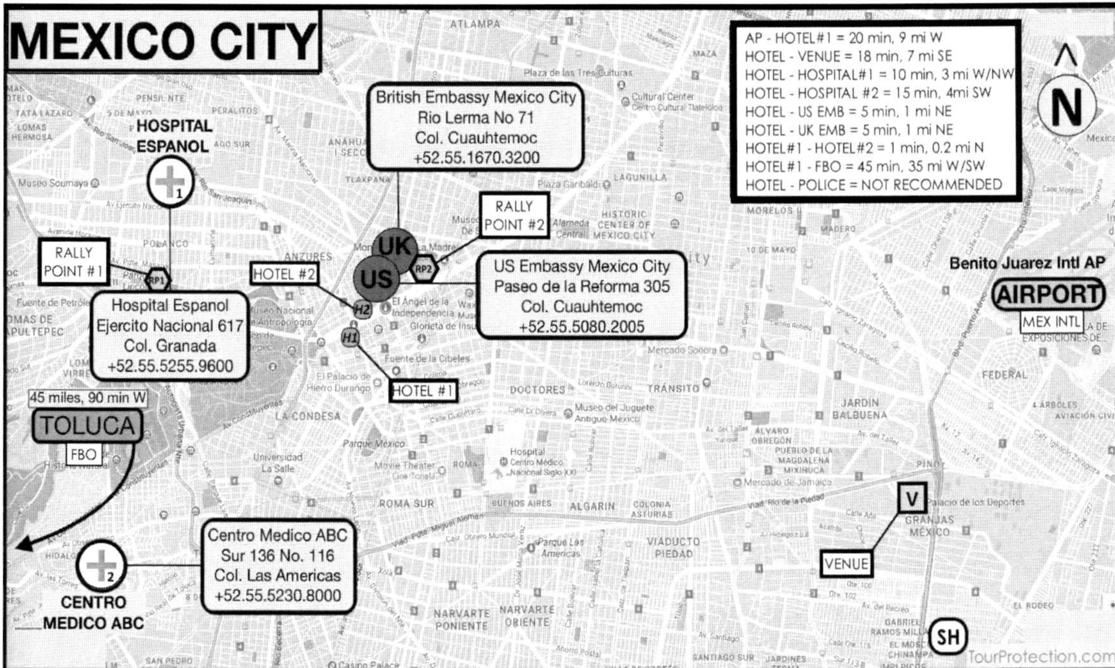
MEXICO CITY
AP - HOTEL#1 = 20 min, 9 mi W
HOTEL - VENUE = 18 min, 7 mi SE
HOTEL - HOSPITAL#1 = 10 min, 3 mi W/NW
HOTEL - HOSPITAL #2 = 15 min, 4mi SW
HOTEL - US EMB = 5 min, 1 mi NE
HOTEL - UK EMB = 5 min, 1 mi NE
HOTEL#1 - HOTEL#2 = 1 min, 0.2 mi N
HOTEL#1 - FBO = 45 min, 35 mi W/SW
HOTEL - POLICE = NOT RECOMMENDED
N
HOSPITAL ESPANOL
British Embassy Mexico City
Rio Lerma No 71
Col. Cuauhtemoc
+52.55.1670.3200
RALLY POINT #2
RALLY POINT #1
HOTEL #2
UK
US
RP1
RP2
H1
H2
US Embassy Mexico City
Paseo de la Reforma 305
Col. Cuauhtemoc
+52.55.5080.2005
Benito Juarez Intl AP
AIRPORT
MEX INTL
Hospital Espanol
Ejercito Nacional 617
Col. Granada
+52.55.5255.9600
HOTEL #1
45 miles, 90 min W
TOLUCA
FBO
V
VENUE
Centro Medico ABC
Sur 136 No. 116
Col. Las Americas
+52.55.5230.8000
CENTRO MEDICO ABC
SH
TourProtection.com

ROUTE SELECTION (BASIC)

Any time you are traveling in a new environment, it's essential to learn about the local roadways, areas they pass through and the pros and cons of your existing route options. Everyone knows that travel can be done by land, air, and sea but which is appropriate? In extreme cases, ground movements are completely avoided and alternative means, such as helicopters are used to avoid roadways.

At a minimum you show know the following:

Know where Danger Zones are located in the area of operation. This typically boils down to the high crime areas in most cases. It can also include areas of different political or religious affiliation.

Know where Choke Points are located on the selected route. The physical design of a city will either help you or hurt you. Understanding how buildings, fields, bases, and other physical elements expose or shield movements lets you identify zones of predictability (ZOPs) and standoff foot prints (SOFs).

Learn your Options along selected routes. Identifying the ZOPs and SOFs will assist with selection and implementation of preventive measures. It will also offer the most likely attacks to occur in that zone. This in turn will allow you to select the most successful counterattack strategy in advance of incident.

Determine Points of No Return along your routes. Sometimes things happen that you could not have predicted. The most common response is to head back to where you came from or a safe haven you have along the route. It's essential to determine a Point of No Return (PONR) for your movement. The PONR will help direct your decision-making process in the event of an incident. If will offer logic and strategy in a time where emotion (not helpful) will run high.

Know Traffic Conditions and Local Activities along the route prior to departure. Traffic is usually, but not always, predictable. Times of days and days of the week often produce the same results within certain parameters. Some events/factors occur outside of those parameters. This may include construction, road closures, check-points, holidays, accidents, parades, protests, or other dignitary operations within that zone. Knowing about them in advance can save you hours. In some cases, it may save a life.

Know Estimated Travel Times/Distances. In order to know what is normal, you need a baseline. When it comes to driving, times are simply estimating. These estimates are used to ensure you are operating according to plan and will keep you on your planned timeline. It will also assist you in identifying an anomaly in your environment that may be a threat, or benign. If you are traveling for business, you can expect your client and others to ask you *"How long will this take?"* on a regular and frequent basis.

Mexico City, MX	11-07-2016
Estimated Times, Dist., Dir.	
Airport to Hotel	20 min, 9 mi W
Hotel to Hospital	10 min, 3 mi NW
Hotel to Embassy	5 min, 1 mi NE
Hotel to Safe Haven	3 min, 1 mi S
Hotel to Venue	18 min, 7 mi E
Hotel to FBO	45 min, 35 mi SW

You can also use this information to offer suggestions such as,

"Ms. Smith, this is a 45-minute drive with no access to restrooms along the route. There is a restroom just across the hallway if you want to use it before we depart."

Know Elements of Nature Around You.
Sunrise, sunset, moon phase, weather conditions, elevation, time zones, and cardinal direction will all help you in planning and preparation that will often equate to survival.

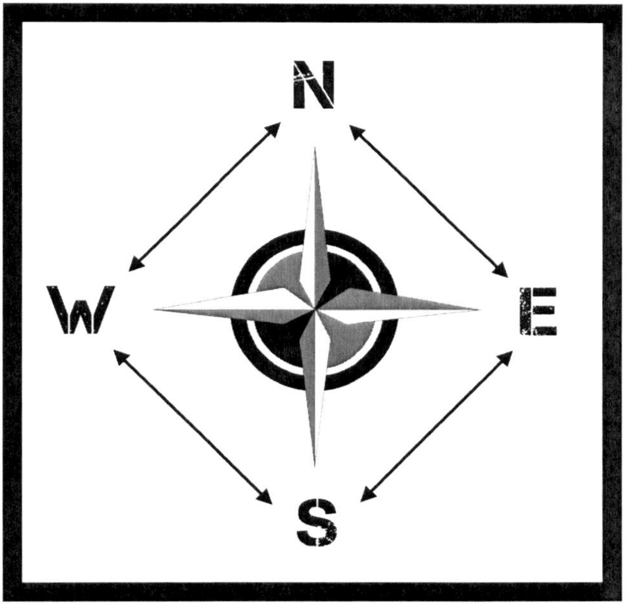
N
W
E
S

Chapter TWENTY-ONE

CONTENT COMPRESSION

PROTECTION FROM
•Physical Harm
•Unintentional Hazards
•Reputational Damage
•Loss of Personal Information
•Anything that could adversely impact protectee

KNOWLEDGE OF
♦Potential problems that can occur in a given environment
♦Possible solutions to the problem
♦Applications required to apply correct solution under stress

CONTACT OPTIONS
♠Best - Don't be where danger commonly occurs
♠Good - Leave as soon as problems start to occur
♠Last/Bad - Stay and fight if you can't leave

OBSERVATION
♠Baseline ID
♠Detect Anomalies
♠Spatial Relations
♠Consider Context
♠No Normalcy Bias
♠Beliefs Drive Actions

ASSESSMENTS
•Criticality
•Accessibility
•Recovery
•Vulnerability
•Effect
•Recognizability

INTELLIGENCE
♣Collect Data
♣Process Data
♣Information
♣Process Info
♣Intelligence
♣Action

PHYSICAL STRUCTURING
♣Deter
♣Delay
♣Defend
♣Respond
♣Mitigate

TERRAIN CONSIDERATIONS
♦Key Terrain
♦Observation & Fields of Fire
♦Cover & Concealment
♦Obstacles/Obstructions
♦Aves of Approach/Escape

HUMAN MOTIVES
♥Money
♥Ideology
♥Coercion
♥Ego

ATTACK EQ.
♠Time
♠Location
♠Method
♠Response

DECISIONS
•Observe
•Orient
•Decide
•Act

"Life is not something, it's an opportunity for something."

-Christian Hebbel

Chapter TWENTY-TWO

MANAGING CHANGE (CHEAT SHEET)

Managing complex change in any organization requires vision, skills, incentives, resources and action plans. If any of these elements are missing, problems will arise. This is well summarized in the image below.

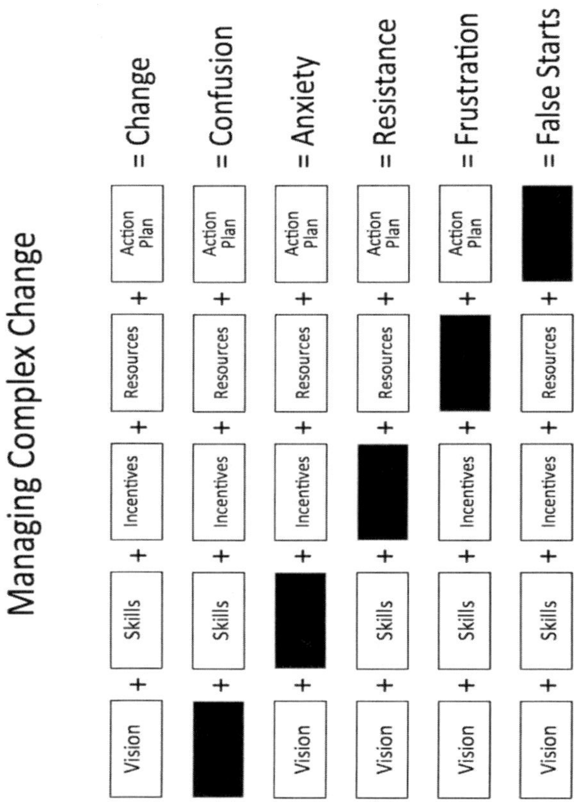

"In theory, there is no difference between theory and practice. In practice, there is."

-Yogi Berra

Chapter TWENTY-THREE

CORE ATTRIBUTES *(OF A PROTECTOR)*

This section would normally be presented on the front end of a book. Due to the dry nature of this section, and technical books in general, the reader was spared, until now. It is essential to understand what areas or attributes allow one to be successful in the protection of someone or something. This may be as a business owner, an employee, or a private citizen looking to secure his family or property.

Awareness The state or ability to perceive, feel, or be conscious of events, objects or patterns relating to the security and safety of a person, place, or thing.

Awareness may be the most essential element of protection. If you are unaware, you have little chance of defeating a threat. Without awareness, you cannot provide protection.

Behavior

The range of actions and mannerisms made in conjunction with their environment. It is the response of the individual to various stimuli or inputs, whether internal/external, conscious/subconscious, overt/covert, and/or voluntary/involuntary.

Behaviors are important indicators that provide advanced warning of a potentially dangerous action. Behaviors often telegraph both intent and actions. Your individual and collective behaviors can encourage or discourage a potential attacker. You must watch others' behaviors as clues while maintaining awareness of your own behaviors. Can you pick up minor indicators of human behavior? If surveillance was being done on you, what would they see and think? How does the person you are protecting behave? Where does that behavior fit? How do *you* behave around the person you protect?

Appearance The way that someone or something looks. An impression given by someone or something.

Humans are acutely sensitive to their physical appearance and that of others. You must take steps to present the proper appearance, which will solicit a desired response or perception from others. How about the appearance of the guy walking toward you? What does his appearance tell you about him? What about your appearance to him? Is it appropriate to the desired perception?

Discretion The quality of behaving or speaking in such a way as to avoid causing offense or revealing private information.

This is essential in the protection of individuals and information, which are often synonymous. Being discreet is the hallmark of a protection professional.

Timing

The choice, judgment, or control of when something should be done. A particular point or period of time when something happens.
Being in the right place at the right time can save your life, while not being in the right place at the right time will often lead to your demise. You cannot depend on others for this. You control your timing. They say timing *is* everything. *It is* in protection.

Suspicious

Having a cautious distrust of something or someone. Understanding that something or someone is (or could be) of questionable, dishonest, or dangerous character or condition.
Pre-mature and misplaced trust often leads to problems of significant value in many realms. In protection, you must have the ability to question motives, statements, and actions to determine if they are logical and/or congruent with other elements of the situation.

Methodical Done according to a systematic or established form of procedure. Orderly in thought and/or behavior.

This assists one from missing or skipping key parts of a protection plan. Plans generally work as a whole and often fail when implemented piecemeal.

Anonymity The state of being anonymous. Unknown and not identified. Having no outstanding, individual, or unusual features; unremarkable or impersonal.

It's difficult for an adversary to plan against you if he doesn't even know you are there. Having anonymity allows you to move freely and not draw unwanted attention to you or the people you are protecting.

Common Sense

You want to develop and hone good sense and sound judgment in practical matters. Unfortunately, this is not always common in today's world.

Communication

It is essential for any individual or team to create and maintain a means of connection between people or places to send and receive information.

Listening Skills

Taking notice of what someone says and responding will keep valuable information flowing in the right direction.

Additional Attributes of Value

Reliability	Consistently good in quality or performance; able to be trusted by others
Punctuality	Doing something at the agreed or proper time
Follow-through	The continuing of an action or task to its conclusion
Privacy	The state or condition of being free from being observed or disturbed by other people
Maturity	The state, fact, or period of being mature
Diplomacy	The art of dealing with people in a sensitive and effective way

A book can teach you. A conversation can assure you. A poem can seduce you. A genius can inspire you. But only YOU can save yourself.

-Anaxagorou

Chapter **TWENTY-FOUR**

CLOSING

Nothing presented in this book is Top Secret, Sensitive Compartmented Information (TS/SCI). It is a collection of basic systems combined, tested, and successfully applied across time and place.

"**Advanced Skills**" are nothing more than properly learned and trained basic skills that are executed well under stress.

This book has offered a basic starting point for understanding of the essential components of ***Protection*** *for + from* ***Humanity***. It is NOT, however, comprehensive or final in nature. It is essential to continue studying, learning, training, applying, and re-assessing. This process never ends. The world is in a constant state of change and that means the mission never ends.

Keep learning.
Keep training.
Keep applying.
Keep re-assessing.
Keep protecting.

"All search for the magic bullet. There are two: SWEAT & FORTITUDE."

-Unknown

I
WON'T
GIVE
UP.

Nothing of value is ever achieved alone. There has always been an abundance of support in the background. The type of support that allows for the success of others. Saying *"thank you"* really doesn't cut it. I am truly grateful to have had access to the humans listed below.

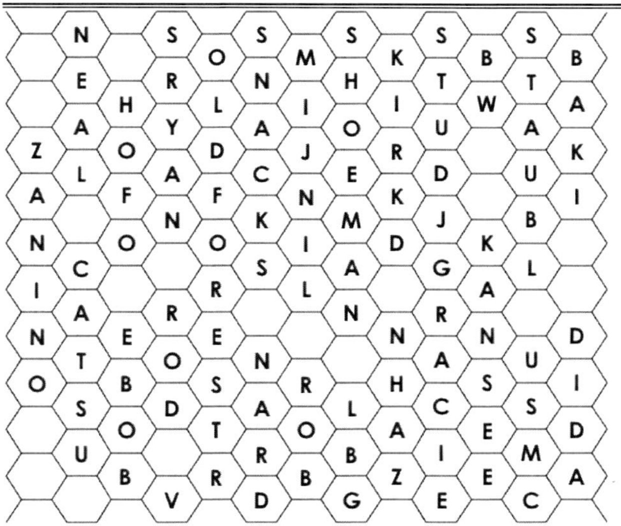

To ALL members of ALL teams I have been a part of. To ALL people who have given me the gift of support.

THANK YOU

"No man steps into the same river twice, for he is not the same man and it is not the same river."

-Heraclitus

ABOUT THE AUTHOR

Todd Fox is the Founder and Managing Director of the Close Protection Corps, a specialized services firm that works with celebrities, dignitaries and C-suite executives in challenging environments. The company was founded in 1999 with their initial services focused on the Latin market.

Mr. Fox's company became known for a number of successful service operations in challenging environments such as Mexico, Colombia, Haiti, Iraq, Turkey and Russia. The company has performed protective operations across the full spectrum of clients ranging from Madonna and Janet Jackson, to Supreme Court Justices and UN Officials.

Mr. Fox is a former U.S. Marine who served honorably for 12-years. He was assigned duties that spanned the Corps' broad gamut. Assignments ranged from washing dishes to operating with JTF-6. His access to "B" billets offered insight into unique realms outside the military. Mr. Fox is also a former law enforcement officer. He was responsible for the creation, development and training of a regional law enforcement special response team (SRT). His team participated in the protection of humans and assets during several pivotal national events. He is a graduate of multiple advanced programs at the Counterterrorism Division (CTD) of the Federal Law Enforcement Training Center (FLETC).

Todd has been a lifelong martial artist. He holds three black belts in uniquely different arts. While in the Marine Corps, he fought in Mixed Martial Arts (MMA) and retired with a professional record of ten wins and two losses. Mr. Fox is a U.S. National Champion, U.S. Open Champion and Pan American Medalist in Brazilian Jiu Jitsu. He continues to actively train in Brazilian Jiu Jitsu and Muay Thai.

He has authored two books on protective service operations and a number of articles as a protective security subject matter expert (SME).

Mr. Fox acts as the Operations Manager for the Fisherman's Fellowship Foundation (F3), a charity founded by his mentor.

Todd Fox is also a lifelong member of the Zanino Group.

CREDITS & WORKS CITED

QUOTES

Anaxagorou A.
Berra, Yogi
Blauer, Tony
Bonaparte, Napoleon
Branson, Sir Richard
Breed, George
Darwin, Charles
Da Vinci, Leonardo
Einstein, Albert
Emerson, Ralph Waldo
Franklin, Benjamin
Goethe
Graham, Gordon
Hannibal
Hawking, Stephen
Hebbel, Christian
Heraclitus
Lawrence, T.E.
McDougall, Chris
Millman, Dan
Plato
Pompeo, Michael
Rogers, Michael
Seneca

WORKS

Boyd, John, *OODA Loop, Decision Cycle, Misc. Presentations (1996)*

Cialdini, Robert, *Presentation, The Power of Persuasion, AADPA Annual Meeting, Palm, CA (2014)*

Cooper, Jeff, *Color Codes & Principles of Protection (2006), Gunsite Academy, Paulden, AZ 86334*

Gordon, Graham, *Presentation, Risk Management (HR/LF Events) to Fire Service, Phoenix, AZ, (2012)*

Klein, Gary, *Sources of Power: How People Make Decisions. Cambridge, MA: MIT Press (1999)*

Sapolsky, Robert, *Behave: The Biology of Humans at Our Best & Worst, Penguin Press (2017)*

United State Marine Corps, *Combat Hunter Program, TBS, MCTC, MCB Quantico, Camp Barrett, VA 22134*

"No man fears that which he knows he does well." -Duke of Wellington